T0399899

Hyers-Ulam Stability of Ordinary Differential Equations

Hyers-Ulam Stability of Ordinary Differential Equations

Arun Kumar Tripathy

CRC Press is an imprint of the
Taylor & Francis Group, an **informa** business
A CHAPMAN & HALL BOOK

First edition published 2021

by CRC Press

6000 Broken Sound Parkway NW, Suite 300, Boca Raton, FL 33487-2742

and by CRC Press

2 Park Square, Milton Park, Abingdon, Oxon, OX14 4RN

CRC Press is an imprint of Taylor & Francis Group, LLC

Library of Congress Cataloging-in-Publication Data

ISBN: [978-0-367-63667-8] (hbk)
ISBN: [978-0-367-63668-5] (pbk)
ISBN: [978-1-003-12017-9] (ebk)

Typeset in CMR10 font
by KnowledgeWorks Global Ltd.

To my family for their endless love, constant support and encouragement.

Contents

Preface

The study of stability problems for various functional equations originated from a talk given by S. M. Ulam in 1940. In that talk Ulam discussed a number of important unsolved problems. Among such problems, a problem concerning the stability of functional equations is one of them. In 1941, D. Hyers gave an answer to the problem on Banach space. Furthermore, the result of Hyers has been generalized by Th. M. Rassias. After that many authors have extended the Ulam's stability problems to other functional equations and generalized Hyer's result in various directions. Thereafter, Ulam's stability problem for functional equations was replaced by stability of differential equations. S. M. Jung has investigated the Hyers-Ulam stability of linear differential equations of different classes. Keeping in view of the definition of Hyers-Ulam stability for differential equations, we can view the corresponding definition of Hyers-Ulam stability for difference equations as well. Because difference equations can be generated from differential equations by using the Euler's method, the study of Hyers-Ulam stability for differential and difference equations is interesting.

The purpose of this book is to reflect the new developments in the Hyers-Ulam Stability Theory for ordinary differential equations and difference equations including some contributions of the author as well as some collections from the published articles. In the proposed topic of interest, first chapter deals with an insight of Hyers-Ulam stability of some functional equations and its origin and subsequent development to ordinary differential and difference equations. In the remaining chapters a discussion has been made comprising of Hyers-Ulam stability of first order, second order, exact and Euler's differential equations. Also, the Hyers-Ulam stability of some complex differential equations is presented. In the last chapter the Hyers-Ulam stability of some difference equations are discussed.

A. K. Tripathy

Author Biography

Dr. Arun Kumar Tripathy, Reader, Department of Mathematics, Sambalpur University, Sambalpur, Odisha, India, is a known name in the literature of Oscillation Theory since past two decades. His contribution basically deals with the linear and nonlinear neutral equations in difference equations, differential equations as well as in time scales of first, second, fourth and higher order equations. It is important that this theory is a part of so-called Dynamical Systems based on Qualitative Behaviour of Solutions Differential and Difference equations. To his credit, he has published 70 research papers in peer-reviewed journals of international repute. Apart from that, he has several international collaborators Prof. S. Pinelas (Portugal), Prof. E. Schmeidal (Poland), Prof. T. G. Baskar (USA), etc. He has been invited to give talks at several international conferences. He is a potential reviewer of many international journals.

After completing successfully the two years Post-Doctoral Fellowship offered by National Board for Higher Mathematics, Department of Atomic Energy, Mumbai, Govt of India, Dr Tripathy started his teaching career in 2003, and till now it is a primary job. Aside from this, research is his special interest, and this interest has been continued since 19 years.

Chapter 1

Introduction and Preliminaries

The subject functional equations generalizes a modern branch of mathematics (see, e.g., [55]). During 1747–1750, J. d'Alembert published three papers and these three papers were the first on functional equations. Many mathematicians like N. H. Abel, J. Bolyai, A. L. Cauchy, J. d'Alembert, L. Euler, M. Frechat, C. F. Gauss, J. L. W. V. Jensen, A. N. Kolmogorov, N. I. Lobacevskii, J. V. Pexider and S. D. Poisson have studied functional equations because of their apparent simplicity and harmonic nature.

In the theory of functional equations we have a question: *When is it true that a function which approximately satisfies a functional equation ϵ must be close to an exact solution ϵ?* If the problem accepts a solution, we say that the equation ϵ is stable.

In 1940, S. M. Ulam [63] has studied the problem for the stability of functional equations in Banach space. Then Ulam questioned that:

Let G_1 be a group and G_2 a metric group with a metric $d(.,.)$. Given $\varepsilon > 0$, does there exist a $\delta > 0$ such that if a mapping $f : G_1 \to G_2$ satisfies $d(f(xy), f(x)f(y)) \leq \delta$ for all $x, y \in G_1$, then there is a homomorphism $g : G_1 \to G_2$ with $d(f(x), g(x)) \leq \varepsilon$ for all $x \in G_1$?

It implies that the homomorphism is stable, i.e. if a mapping is almost a homomorphism, then there exists a true homomorphism near it.

A certain formula or equation is applicable for solution of a problem. A small change in the formula or equation gives rise to a small change in the corresponding result. In this case we call this formula or equation is stable. The quadratic equation

$$f(x+y) = f(x) + f(y)$$

is not always true for all $x, y \in \mathbb{R}$ but it may be true approximately, that is,

$$f(x+y) - f(x) - f(y) \approx 0,$$

for all $x, y \in \mathbb{R}$. In particular, if we choose $f(x) = sinx$, then the above equation doesn't hold. This can be stated mathematically as

$$|f(x+y) - f(x) - f(y)| \leq \varepsilon$$

for small positive ε and for all $x, y \in \mathbb{R}$. This is called additive Cauchy functional equation. If there is a small change on Cauchy additive functional equation, we take only small effects on its solution. This is called the Stability theory.

Definition 1 *Let E_1 be a group and E_2 a quasi-normed space. If the functions $F, G : E_1 \to E_2$ fulfil the inequality*

$$\begin{aligned}&d[F(x+y+z)+F(x-y)+F(y-z)+F(x-z),\\&G(x)+G(y)+G(z)] \leq h(x,y,z)\end{aligned}$$

and h is constant, then it is called Hyers-Ulam Stability. Rassias introduced the inequality of the type

$$\|f(x+y)-f(x)-f(y)\| \leq \theta(\|x\|^p+\|y\|^p), \;\; \theta \geq 0$$

is called Hyers-Ulam-Rassias stability (*generalized Hyers-Ulam stability*). *If the inequality*

$$\|f(x+y)-f(x)-f(y)\| \leq \varepsilon,$$

is replaced by

$$\|f(x+y)-f(x)-f(y)\| \leq \varepsilon(\|x\|^p+\|y\|^p), \;\; \varepsilon > 0,$$

it is called generalized Hyers-Ulam-Rassias stability.

Definition 2 *An additive mapping or additive function that preserves the addition operation if*

$$f(x+y) = f(x)+f(y).$$

Definition 3 *For a given mapping $f : X \to Y$ is called additive if it satisfies*

$$f(rx+sy) = \frac{r+s}{2}f(x+y)+\frac{r-s}{2}f(x-y)$$

$r, s \in \mathbb{R}$ and $r \neq s$.

Definition 4 *Let A be a Banach algebra. An algebraic left A-module X is said to be a normed (Banach) left A-module if X is a normed (Banach) space and the outer multiplication is jointly continuous, that is, if there is a non-negative number M such that $\|ax\| \leq M\|a\|\|x\|$, $a \in A$, $x \in X$. If A has an identity e, then X is called unital if $ex = x$ for all $x \in X$.*
For example, every closed left ideal I of a normed algebra A can be regarded as a Banach left A-module with the product of A giving the module multiplication.

Definition 5 *A C^* algebra is a complex algebra A of continuous linear operator on a complex Hilbert space with following two properties :*
1. A is a topologically closed set in the norm topology of operators.
2.A is closed under the operation of taking adjoints of operators.

Definition 6 *Given two groups $(G, *)$ and $(H, .)$, a group homomorphism from $(G, *)$ to $(H, .)$ is a function $h : G \to H$ such that for all u and v in G it holds that*

$$h(u * v) = h(u).h(v),$$

where the group operation on the left hand side of the equations is that of G and on the right hand side that of H.

Definition 7 *An additive mapping $f : X \to Y$ is called A-algebra if $f(ax) = af(x)$ for all $x \in X$ and all $a \in A$. Similarly if $a \in \mathbb{C}$, then it is called $\mathbb{C}$-linear.*

Theorem 1 *Let (X, d) be a complete generalized metric space and let $J : X \to X$ be a strictly contractive mapping with Lipschitz constant $L < 1$. Then for each element $x \in X$, either*

$$d(J^n x, J^{n+1} x) = \infty,$$

for all nonnegative integers n or there exists a positive integer n_0 such that

(1) $d(J^n x, J^{n+1} x) < \infty$, *for all* $n \geq n_0$;

(2) *the sequence* $\{J^n x\}$ *converges to a fixed point* y^* *of* J;

(3) y^* *is the unique fixed point of* J *in the set* $Y = \{y \in X | d(J^{n_0} x, y) < \infty\}$;

(4) $d(y, y^*) \leq \frac{1}{1-L} d(y, Jy)$ *for all* $y \in Y$.

Theorem 2 *Let E be a (real or complex) linear space, E and F are Banach spaces, and*

$$q_i = \begin{cases} 2, & i = 0 \\ 1/2, & i = 1. \end{cases}$$

Suppose that the mapping $f : E \to F$ satisfies the condition $f(0) = 0$ and an inequality of the form
(J_φ) $\left\|2f\left(\frac{x+y}{2}\right) - f(x) - f(y)\right\|_F \leq \varphi(x, y)$,
for all $x, y \in E$, where $\varphi : E \times E \to [0, \infty)$ is a given function. If there exists $L = L(i) < 1$ such that the mapping $x \to \psi(x) = \varphi(x, 0)$ has the property
(H_i) $\psi(x) \leq L.q_i.\psi(x/q_i)$ *for all* $x \in E$
and the mapping φ has property
(H_{i^*}) $\lim_{n\to\infty} \varphi(2q_i^n x, 2q_i^n y) = 0$ *for all* $x, y \in E$
then there exists a unique additive mapping $j : E \to F$ such that

$$(E_{sti})\ \|f(x) - j(x)\|_F \leq \frac{L^{1-i}}{1-L}\psi(x)$$

for all $x \in E$.

1.1 Stability of Functional Equations on Banach Space

Ulam proposed the problem: *When does a linear transformation near an "approximately linear" transformation exist?*

This problem was solved partially by D. H. Hyers on Banach spaces. Let E and E' be Banach spaces. Then a transformation $f(x)$ from E into E' is called approximately linear if there exist $K \geq 0$ and $p(0 \leq p < 1)$ such that

$$\|f(x+y) - f(x) - f(y)\| \leq K(\|x\|^p + \|y\|^p) \quad for \quad any \quad x, y \in E.$$

If there exist two transformations $f(x)$ and $\varphi(x)$ from E into E' and $K \geq 0$ and $p\,(0 \leq p < 1)$ such that

$$\|f(x) - \varphi(x)\| \leq K\|x\|^p,$$

then these are called near.

1.1.1 Stability of $f(x+y) = f(x) + f(y)$

Theorem 3 *If $f(x)$ is an approximately linear transformation from E into E' then there is a linear transformation $\varphi(x)$ near f(x) and such $\varphi(x)$ is unique.*

Proof 1 *Let us assume that $K_0 \geq 0$ and $p(0 \leq p < 1)$ such that*

$$\|f(x+y) - f(x) - f(y)\| \leq K_0(\|x\|^p + \|y\|^p). \tag{1.1}$$

Putting $x = y$ in (1.1), we get

$$\|f(2x) - 2f(x)\| \leq 2K_0(\|x\|^p),$$

that is,

$$\|f(2x)/2 - f(x)\| \leq K_0\|x\|^p. \tag{1.2}$$

We claim that

$$\|f(2^n x)/2^n - f(x)\| \leq K_0\|x\|^p \sum_{i=0}^{n-1} 2^{i(p-1)}, \quad n \in N. \tag{1.3}$$

Clearly $n = 1$ holds by (1.2). Assume that (1.3) holds for n. We shall show that (1.3) holds for $n = n+1$. Replacing x by $2x$ in (1.3), we obtain

$$\begin{aligned}\|f(2^{n+1}x)/2^{n+1} - f(2x)\| &\leq K_0\|2x\|^p \sum_{i=0}^{n-1} 2^{i(p-1)} \\ &= K_0\|x\|^p 2^p \sum_{i=0}^{n-1} 2^{i(p-1)}.\end{aligned}$$

Therefore,

$$\begin{aligned}\|f(2^{n+1}x)/2^{n+1} - f(2x)/2\| &\leq K_0\|x\|^p 2^{p-1} \sum_{i=0}^{n-1} 2^{i(p-1)} \\ &= K_0\|x\|^p \sum_{i=1}^{n} 2^{i(p-1)}.\end{aligned}$$

Now,

$$\begin{aligned}\|f(2^{n+1}x)/2^{n+1} - f(x)\| &\leq \|f(2^{n+1}x)/2^{n+1} - f(2x)/2\| + \|f(2x)/2 - f(x)\| \\ &\leq K_0\|x\|^p[1 + \sum_{i=1}^{n} 2^{i(p-1)}] \\ &= K_0\|x\|^p \sum_{i=0}^{n} 2^{i(p-1)}\end{aligned}$$

implies that (1.3) is true for all n. *Indeed,*

$$\begin{aligned}\|f(2^n x)/2^n - f(x)\| &\leq K_0\|x\|^p \sum_{i=0}^{n-1} 2^{i(p-1)} \\ &\leq K_0\|x\|^p \sum_{i=0}^{\infty} 2^{i(p-1)} \\ &\leq K\|x\|^p 2/2 - 2^p = K\|x\|^p,\end{aligned}$$

where $K = 2K_0/2 - 2^p$.
Next, we claim that $\{f(2^n x)/2^n\}$ *is a Cauchy sequence. Therefore,*

$$\begin{aligned}\|f(2^m x)/2^m - f(2^n x)/2^n\| &= \frac{1}{2^n}\|f(2^m x)/2^{m-n} - f(2^n x)\| \\ &= \frac{1}{2^n}\|f(2^{m-n}2^n x)/2^{m-n} - f(2^n x)\| \\ &\leq \frac{K}{2^n}\|2^n x\|^p \\ &= 2^{n(p-1)}K\|x\|^p \quad \to 0 (n \to \infty).\end{aligned}$$

So, $\{f(2^n x)/2^n\}$ *is a Cauchy sequence. Since* E' *is complete, then* $\{f(2^n x)/2^n\}$ *converges. Let*

$$\varphi(x) \equiv \lim_{n\to\infty} f(2^n x)/2^n.$$

We assert that $\varphi(x)$ *is linear. Since* $f(x)$ *is approximate linear, then*

$$\begin{aligned}\|f(2^n(x+y)) - f(2^n x) - f(2^n y)\| &\leq K_0(\|2^n x\|^p + \|2^n y\|^p) \\ &\leq 2^{np}K_0(\|x\|^p + \|y\|^p),\end{aligned}$$

that is,

$$\|f(2^n(x+y))/2^n - f(2^n x)/2^n - f(2^n y)/2^n\| \leq 2^{n(p-1)}K_0(\|x\|^p + \|y\|^p).$$

Taking limit as $n \to \infty$, *we get*

$$\varphi(x+y) = \varphi(x) + \varphi(y).$$

So, φ is linear. Taking $\lim_{n\to\infty}$ in (1.3), we find

$$\|\varphi(x) - f(x)\| \leq K\|x\|^p$$

which shows that $\varphi(x)$ is near $f(x)$.

Finally, we have to show that $\varphi(x)$ is unique. If not, then there exists $\psi(x)$ near $f(x)$ such that for $K'(\geq 0)$, $0 \leq p' < 1$, we have

$$\|\psi(x) - f(x)\| \leq K'\|x\|^{p'}.$$

By the triangle inequality

$$\begin{aligned}\|\varphi(x) - \psi(x)\| &\leq \|\varphi(x) - f(x)\| + \|\psi(x) - f(x)\| \\ &\leq K\|x\|^p + K'\|x\|^{p'}.\end{aligned}$$

Since φ and ψ are linear, then

$$\begin{aligned}\|\varphi(x) - \psi(x)\| &= \|\varphi(nx) - \psi(nx)\|/n \\ &= (K\|nx\|^p + K'\|nx\|^{p'})/n \\ &= K\|x\|^p/n^{1-p} + K'\|x\|^{p'}/n^{1-p'} \\ &\to 0 \ \text{as} \ n \to \infty.\end{aligned}$$

This completes the proof of the theorem.

In the above Theorem 3, we proved the Ulam's problem. Now we prove a generalization of the stability of approximately additive mappings in the spirit of Hyers-Ulam and Rassias.
Denote $(G, +)$ an abelian group, $(X, \|.\|)$ a Banach space and $\varphi : G \times G \to [0, \infty)$ be a mapping such that

$$\widetilde{\varphi}(x, y) = \sum_{k=0}^{\infty} 2^{-k}\varphi(2^k x, 2^k y) < \infty \quad \text{for all } x, y \in G. \tag{1.4}$$

1.1.2 Stability of $f(x + y) = f(x) + f(y)$

Theorem 4 *Let $f : G \to X$ be such that*

$$\|f(x + y) - f(x) - f(y)\| \leq \varphi(x, y) \quad \text{for all } x, y \in G. \tag{1.5}$$

Then there exists a unique mapping $T : G \to X$ such that

$$T(x + y) = T(x) + T(y), \quad \text{for all } x, y \in G. \tag{1.6}$$

and

$$\|f(x) - T(x)\| \leq \frac{1}{2}\widetilde{\varphi}(x, x) \quad \text{for all } x \in G. \tag{1.7}$$

Proof 2 *Putting $x = y$ in (1.5), we get*

$$\|f(2x) - 2f(x)\| \leq \varphi(x, x).$$

Thus,

$$\|2^{-1}f(2x) - f(x)\| \leq \frac{1}{2}\varphi(x, x) \quad \text{for all } x \in G. \tag{1.8}$$

Replacing x by $2x$ in (1.8), we get

$$\|2^{-1}f(2^2x) - f(2x)\| \leq \frac{1}{2}\varphi(2x, 2x) \quad \text{for all } x \in G. \tag{1.9}$$

Hence,

$$\begin{aligned}\|2^{-2}f(2^2x) - f(x)\| &\leq \|2^{-2}f(2^2x) - 2^{-1}f(2x)\| \\ &\quad + \|2^{-1}f(2x) - f(x)\| \\ &\leq 2^{-1}\|2^{-1}f(2^2x) - f(2x)\| + \|2^{-1}f(2x) - f(x)\| \\ &\leq 2^{-1}\frac{1}{2}\varphi(2x, 2x) + \frac{1}{2}\varphi(x, x). \end{aligned} \tag{1.10}$$

Replacing x by $2x$ in (1.10), we find

$$\|2^{-2}f(2^3x) - f(2x)\| \leq \frac{1}{2}[\varphi(2x, 2x) + \frac{1}{2}\varphi(2^2x, 2^2x)],$$

and therefore

$$\begin{aligned}\|2^{-3}f(2^3x) - f(x)\| &\leq \|2^{-3}f(2^3x) - 2^{-1}f(2x)\| \\ &\quad + \|2^{-1}f(2x) - f(x)\| \\ &\leq 2^{-1}\frac{1}{2}[\varphi(2x, 2x) + \frac{1}{2}\varphi(2^2x, 2^2x)] + \frac{1}{2}\varphi(x, x). \end{aligned} \tag{1.11}$$

Proceeding inductively, we obtain

$$\|2^{-n}f(2^nx) - f(x)\| \leq \frac{1}{2}\sum_{k=0}^{n-1} 2^{-k}\varphi(2^kx, 2^kx) \quad \text{for all } x \in G. \tag{1.12}$$

The process will be complete if, we show that (1.12) holds for $n = n + 1$. Clearly,

$$\begin{aligned}\|2^{-(n+1)}f(2^{n+1}x) - f(x)\| &\leq \|2^{-(n+1)}f(2^{n+1}x) - 2^{-1}f(2x)\| \\ &\quad + \|2^{-1}f(2x) - f(x)\| \\ &\leq 2^{-1}\|2^{-n}f(2^{n+1}x) - f(2x)\| + \|2^{-1}f(2x) - f(x)\| \\ &\leq 2^{-1}\frac{1}{2}\sum_{k=0}^{n-1} 2^{-k}\varphi(2^{k+1}x, 2^{k+1}x) + \frac{1}{2}\varphi(x, x) \\ &= \frac{1}{2}\sum_{k=0}^{n} 2^{-k}\varphi(2^kx, 2^kx). \end{aligned}$$

We claim that $\{2^{-n}f(2^n x)\}$ *is a Cauchy sequence. For* $n > m$*, we have*

$$\begin{aligned}\|2^{-n}f(2^n x) - 2^{-m}f(2^m x)\| &= 2^{-m}\|2^{-(n-m)}f(2^{n-m}2^m x) - f(2^m x)\| \\ &\le 2^{-m}\frac{1}{2}\sum_{k=0}^{n-m-1} 2^{-k}\varphi(2^{k+m}x, 2^{k+m}x) \\ &\le \frac{1}{2}\sum_{k=0}^{n-m-1} 2^{-(k+m)}\varphi(2^{k+m}x, 2^{k+m}x) \\ &= \frac{1}{2}\sum_{p=m}^{n-1} 2^{-p}\varphi(2^p x, 2^p x),\end{aligned}$$

where $p = k + m$*. When* $k = 0, p = m$ *and* $k = n - m - 1, p = n - 1$*. Taking limit as* $n \to \infty$ *in the above result and keeping* m *as fixed, we obtain*

$$\lim_{n\to\infty} \|2^{-n}f(2^n x) - 2^{-m}f(2^m x)\| = 0.$$

Since X *is a Banach space, then it follows that the sequence* $\{2^{-n}f(2^n x)\}$ *converges in* X*. Let*

$$T(x) = \lim_{n\to\infty} \frac{f(2^n x)}{2^n}.$$

We claim that T *satisfies* (1.6)*. From (1.5), we have*

$$\|f(2^n x + 2^n y) - f(2^n x) - f(2^n y)\| \le \varphi(2^n x, 2^n y) \quad \textit{for all} \quad x, y \in G$$

and

$$\|2^{-n}f(2^n x + 2^n y) - 2^{-n}f(2^n x) - 2^{-n}f(2^n y)\| \le 2^{-n}\varphi(2^n x, 2^n y).$$

Therefore,

$$\lim_{n\to\infty} \|2^{-n}f(2^n x + 2^n y) - 2^{-n}f(2^n x) - 2^{-n}f(2^n y)\| \le \lim_{n\to\infty} 2^{-n}\varphi(2^n x, 2^n y) \tag{1.13}$$

implies that

$$\|T(x + y) - T(x) - T(y)\| = 0$$

due to (1.4)*. Taking limit as* $n \to \infty$ *in* (1.12)*, we obtain*

$$\|T(x) - f(x)\| \le \frac{1}{2}\tilde{\varphi}(x, x) \quad \textit{for all} \quad x \in G.$$

Next, we show that T *is unique. If not, then there exists* $F : G \to X$ *with*

$$F(x + y) = F(x) + F(y)$$

and satisfies (1.7). *Then,*

$$\begin{aligned}\|T(x)-F(x)\| &= \|2^{-n}T(2^nx)-2^{-n}F(2^nx)\| \\ &\leq \|2^{-n}T(2^nx)-2^{-n}f(2^nx)\|+\|2^{-n}f(2^nx)-2^{-n}F(2^nx)\| \\ &\leq 2^{-n}\frac{1}{2}\widetilde{\varphi}(2^nx,2^nx)+2^{-n}\frac{1}{2}\widetilde{\varphi}(2^nx,2^nx) \\ &= 2^{-n}\widetilde{\varphi}(2^nx,2^nx) \\ &= 2^{-n}\sum_{k=0}^{\infty}2^{-k}\varphi(2^{k+n}x,2^{k+n}x) \\ &= \sum_{p=n}^{\infty}2^{-p}\varphi(2^px,2^px).\end{aligned}$$

Hence,

$$\|T(x)-F(x)\| \leq \sum_{p=n}^{\infty}2^{-p}\varphi(2^px,2^px) \ \ for \ \ all \ \ x\in G. \tag{1.14}$$

Taking limit as $n\to\infty$ *in* (1.14), *we obtain*

$$T(x)=F(x) \ \ for \ \ all \ \ x\in G.$$

Example 1 *Let* G *be a normed linear space and define* $H: R_+\times R_+\to R_+$ *and* $\varphi_0: R_+\to R_+$ *such that*

$$\varphi_0(\lambda)>0 \ \ for \ all \ \lambda>0.$$

$$\varphi_0(2)<2,$$

$$\varphi_0(2\lambda)\leq\varphi_0(2)\varphi_0(\lambda) \ \ for \ all \ \lambda>0.$$

$$H(\lambda t,\lambda s)\leq\varphi_0(\lambda)H(t,s) \ \ for \ \ all \ \ t,s\in R_+,\lambda>0.$$

In Theorem 4, let $\varphi(x,y)=H(\|x\|,\|y\|)$. *Then,*

$$\begin{aligned}\varphi(2^kx,2^ky) &= H(2^k\|x\|,2^k\|\|y\|) \\ &\leq \varphi_0(2^k)H(\|x\|,\|y\|) \\ &\leq [\varphi_0(2)\varphi_0(2)\varphi_0(2)\cdots k-times]H(\|x\|,\|y\|) \\ &\leq (\varphi_0(2))^kH(\|x\|,\|y\|).\end{aligned}$$

But, $\varphi_0(2)<2$. *So,*

$$\begin{aligned}\widetilde{\varphi}(x,y) &= \sum_{k=0}^{\infty}2^{-k}\varphi(2^kx,2^ky) \\ &\leq \sum_{k=0}^{\infty}2^{-k}(\varphi_0(2))^kH(\|x\|,\|y\|) \\ &= \sum_{k=0}^{\infty}(\varphi_0(2)/2)^kH(\|x\|,\|y\|) \\ &= \frac{1}{1-(\varphi_0(2)/2)}H(\|x\|,\|y\|).\end{aligned}$$

Therefore, $\|f(x) - T(x)\| \leq \frac{1}{2}\widetilde{\varphi}(x,x)$ *for all* $x \in G$ *implies that*

$$\begin{aligned}\|f(x) - T(x)\| &\leq \frac{1}{2-\varphi_0(2)} H(\|x\|, \|x\|)\\ &\leq \frac{1}{2-\varphi_0(2)} \varphi_0(\|x\|) H(1,1).\end{aligned}$$

Remark 1 *The above result generalizes that if* $f(tx)$ *is continuous in* t *for each fixed* x *and*

$$T(x) = \lim_{n\to\infty} 2^{-n} f(2^n x),$$

then T *is a linear mapping.*

1.1.3 Stability of $f(x + y) = g(x) + h(y)$

Let $(X, +)$, $(Y, +)$ be abelian groups and f, g, h: $X \to Y$ be mappings. If f, g and h satisfy the functional equation

$$f(x + y) - g(x) - h(y) = 0 \quad for \ all \ x, y \in X,$$

then it is called as Pexider equation.

Let $(G, +)$ be an abelian group and $(X, \|.\|)$ be a Banach space.We define $\varphi : G \times G \to [0, \infty)$ be such that

$$\varepsilon(x) = \sum_{j=1}^{\infty} \frac{\varphi(2^{j-1}x, 0) + \varphi(0, 2^{j-1}x) + \varphi(2^{j-1}x, 2^{j-1}x)}{2^j} < \infty, \tag{1.15}$$

and

$$\frac{\phi(2^n x, 2^n y)}{2^n} \to 0 \ as \ n \to \infty \ for \ all \ x, y \in G.$$

Theorem 5 *Let f, g, h:* $G \to X$ *be mappings satisfying the inequality*

$$\|f(x + y) - g(x) - g(y)\| \leq \varphi(x, y) \tag{1.16}$$

for all $x, y \in G$. *Then there exists a unique additive mapping* $T : G \to X$ *such that*

$$\|f(x) - T(x)\| \leq \|g(0)\| + \|h(0)\| + \varepsilon(x), \tag{1.17}$$

$$\|g(x) - T(x)\| \leq \|g(0)\| + 2\|h(0)\| + \varphi(x, 0) + \varepsilon(x), \tag{1.18}$$

$$\|h(x) - T(x)\| \leq 2\|g(0)\| + \|h(0)\| + \varphi(0, x) + \varepsilon(x) \tag{1.19}$$

Proof 3 *Putting* $x = y$ *in (1.16), we find*

$$\|f(2x) - g(x) - h(x)\| \leq \varphi(x, x) \tag{1.20}$$

for all $x \in G$. *For* $y = 0$ *in (1.16), we get*

$$\|f(x) - g(x) - h(0)\| \leq \varphi(x, 0) \quad for \ all \ x \in G. \tag{1.21}$$

From (1.21), it follows that

$$\|g(x)-f(x)\|-\|h(0)\| \leq \varphi(x,0) \;\; for \;\; all \;\; x \in G. \tag{1.22}$$

Let $x=0$ *in (1.16). Then,*

$$\|f(y)-g(0)-h(y)\| \leq \varphi(0,y) \;\; for \;\; all \;\; y \in G. \tag{1.23}$$

Consequently, (1.23) becomes

$$\|h(x)-f(x)\| \leq \|g(0)\|+\varphi(0,x) \;\; for \;\; all \;\; x \in G. \tag{1.24}$$

Assume that

$$u(x) := \|h((0)\|+\|g(0)\|+\varphi(0,x)+\varphi(x,0)+\varphi(x,x), \;\; x \in G.$$

Since,

$$\begin{aligned}\|f(2x)-2f(x)\| \leq \|f(2x)-g(x)-h(x)\|+\|g(x)-f(x)\| \\ +\|h(x)-f(x)\|,\end{aligned}$$

then using (1.20),(1.22) and (1.24), we obtain

$$\begin{aligned}\|f(2x)-2f(x)\| &\leq \|h(0)\|+\|g(0)\|+\varphi(0,x)+\varphi(x,0)+\varphi(x,x) \\ &= u(x) \;\; for \;\; all \;\; x \in G. \end{aligned} \tag{1.25}$$

Replacing x *by* $2x$ *in (1.25), it happens that*

$$\|f(2^2x)-2f(2x)\| \leq u(2x) \;\; for \;\; all \;\; x \in G \tag{1.26}$$

and hence

$$\begin{aligned}\|f(2^2x)-2^2f(x)\| &\leq \|f(2^2x)-2f(2x)\|+2\|f(2x)-2f(x)\| \\ &\leq u(2x)+2u(x) \;\; for \;\; all \;\; x \in G. \end{aligned} \tag{1.27}$$

We claim that

$$\|f(2^nx)-2^nf(x)\| \leq \sum_{j=1}^{n} 2^{j-1}u(2^{n-j}x) \;\; for \;\; all \;\; x \in G. \tag{1.28}$$

By the method of induction, (1.28) is true for $n=1$ *due to (1.25). Substituting* $2x$ *for* x *in (1.28), we obtain*

$$\|f(2^{n+1}x)-2^nf(2x)\| \leq \sum_{j=1}^{n} 2^{j-1}u(2^{n+1-j}x) \;\; for \;\; all \;\; x \in G. \tag{1.29}$$

Hence, using (1.29)

$$\begin{aligned}\|f(2^{n+1}x) - 2^{n+1}f(x)\| &\le \|f(2^{n+1}x) - 2^n f(2x)\| + 2^n\|f(2x) - 2f(x)\| \\ &\le \sum_{j=1}^{n} 2^{j-1}u(2^{n+1-j}x) + 2^n u(x) \\ &= \sum_{j=1}^{n+1} 2^{j-1}u(2^{n+1-j}x) \end{aligned} \tag{1.30}$$

for all $x \in G$. So, our claim is true. It follows from (1.28) that

$$\|2^{-n}f(2^n x) - f(x)\| \le \sum_{j=1}^{n} 2^{j-1-n}u(2^{n-j}x) \quad \textit{for all} \quad x \in G. \tag{1.31}$$

We show that $\{2^{-n}f(2^n x)\}$ is a Cauchy sequence in X. For $m < n$,

$$\begin{aligned}\|2^{-n}f(2^n x) - 2^{-m}f(2^m x)\| &= \|2^{-n}f(2^n x) - 2^{-m}f(2^m x) + 2^{-(m+1)}f(2^{m+1}x) \\ &\quad - 2^{-(m+1)}f(2^{m+1}x) + \cdots + 2^{-(n-1)}f(2^{n-1}x) - 2^{-(n-1)}f(2^{n-1}x)\| \\ &\le \|2^{-m}f(2^m x) - 2^{-(m+1)}f(2^{m+1}x)\| + \cdots \\ &\quad + 2^{-(n-1)}f(2^{n-1}x) - 2^{-n}f(2^n x)\| \\ &\le \sum_{j=m}^{n-1} \|2^{-j}f(2^j x) - 2^{-(j+1)}f(2^{j+1}x)\| \\ &\le \sum_{j=m}^{n-1} 2^{-(j+1)}\|f(2^{j+1}x) - 2f(2^j x)\| \\ &\le \sum_{j=m}^{n-1} 2^{-(j+1)}u(2^j x) \\ &\le \sum_{j=m}^{n-1} \frac{u(2^j x)}{2^{j+1}} \\ &= \sum_{j=m}^{n-1} \frac{\|g(0)\| + \|h(0)\| + \varphi(2^j x, 0) + \varphi(0, 2^j x) + \varphi(2^j x, 2^j x)}{2^{j+1}} \\ &= \sum_{j=m}^{n-1} \frac{\|g(0)\| + \|h(0)\|}{2^{j+1}} \\ &\quad + \sum_{j=m}^{n-1} \frac{\varphi(2^j x, 0) + \varphi(0, 2^j x) + \varphi(2^j x, 2^j x)}{2^{j+1}} \\ &= \|g(0)\| + \|h(0)\|(1/2^{m+1} + 1/2^{m+2} + \cdots + 1/2^n) \\ &\quad + \sum_{j=m}^{n-1} \frac{\varphi(2^j x, 0) + \varphi(0, 2^j x) + \varphi(2^j x, 2^j x)}{2^{j+1}}\end{aligned}$$

$$= \frac{\|g(0)\| + \|h(0)\|}{2^{m+1}}(1 + 1/2 + 1/2^2 + \cdots + 1/2^{n-m-1})$$
$$+ \sum_{j=m}^{n-1} \frac{\varphi(2^j x, 0) + \varphi(0, 2^j x) + \varphi(2^j x, 2^j x)}{2^{j+1}}$$

Since $2^{m+1} > 2^m$, *then*

$$\|2^{-n} f(2^n x) - 2^{-m} f(2^m x)\| \le \frac{\|g(0)\| + \|h(0)\|}{2^m} + \sum_{j=1}^{\infty} \frac{\varphi(2^j x, 0) + \varphi(0, 2^j x) + \varphi(2^j x, 2^j x)}{2^{j+1}} \tag{1.32}$$

for all $x \in G$. *Taking limit as* $n \to \infty$, *and keeping* m *as fixed, we find*

$$\lim_{n\to\infty} \|2^{-n} f(2^n x) - 2^{-m} f(2^m x)\| = 0$$

for all $x \in G$. *Since* X *is a Banach space, then it follows that the sequence* $\{2^{-n} f(2^n x)\}$ *converges. Define* $T : G \to X$ *by*

$$T(x) = \lim_{n\to\infty} 2^{-n} f(2^n x).$$

We claim that T *satisfies (1.17). From (1.16), we have*

$$\|\frac{f(2^n x + 2^n y)}{2^n} - \frac{g(2^n x)}{2^n} - \frac{h(2^n y)}{2^n}\| \le \frac{\varphi(2^n x, 2^n y)}{2^n} \tag{1.33}$$

for all $x, y \in G$. *From(1.22) we get,*

$$\|\frac{g(2^n x)}{2^n} - \frac{f(2^n x)}{2^n}\| \le \frac{\|h(0)\| + \varphi(2^n x, 0)}{2^n} \tag{1.34}$$

for all $x \in G$. *Since*

$$\frac{\varphi(2^n x, 0)}{2^n} \le 2 \sum_{j=n}^{\infty} \frac{\varphi(2^j x, 0) + \varphi(0, 2^j x) + \varphi(2^j x, 2^j x)}{2^{j+1}} \tag{1.35}$$

$$\to 0 \text{ as } n \to \infty,$$

then we obtain from (1.34) that

$$\lim_{n\to\infty} \frac{g(2^n x)}{2^n} = \lim_{n\to\infty} \frac{f(2^n x)}{2^n} \tag{1.36}$$

for all $x \in G$. *Therefore, from (1.24)*

$$\|\frac{h(2^n x)}{2^n} - \frac{f(2^n x)}{2^n}\| \le \frac{\|g(0)\| + \varphi(0, 2^n x)}{2^n} \tag{1.37}$$

for all $x \in G$ and hence

$$\lim_{n\to\infty} \frac{h(2^n x)}{2^n} = \lim_{n\to\infty} \frac{f(2^n x)}{2^n} \tag{1.38}$$

for all $x \in G$. Due to (1.36), (1.38), and G is commutative, we conclude that

$$\begin{aligned} 0 &= \left\| \lim_{n\to\infty} \left(\frac{f(2^n x + 2^n y)}{2^n} - \frac{g(2^n x)}{2^n} - \frac{h(2^n y)}{2^n} \right) \right\| \\ &= \|T(x+y) - \lim_{n\to\infty} \frac{f(2^n x)}{2^n} - \lim_{n\to\infty} \frac{f(2^n y)}{2^n} \| \\ &= \|T(x+y) - T(x) - T(y)\| \ \ for \ \ all \ \ x, y \in G. \end{aligned} \tag{1.39}$$

Taking limit as $n \to \infty$ in (1.31), we have

$$\begin{aligned} \|T(x) - f(x)\| &\le \lim_{n\to\infty} \sum_{j=1}^{n} 2^{j-1-n} u(2^{n-j} x) \\ &= \lim_{n\to\infty} \sum_{j=1}^{n} 2^{j-1-n} [\|g(0)\| + \|h(0)\| + \varphi(0, 2^{n-j} x) \\ &+ \varphi(2^{n-j} x, 0) + \varphi(2^{n-j} x, 2^{n-j} x)] \\ &= \lim_{n\to\infty} \sum_{j=1}^{n} 2^{j-1-n} (\|g(0)\| + \|h(0)\|) \\ &+ \lim_{n\to\infty} \sum_{j=1}^{n} \frac{\varphi(2^{j-1} x, 0) + \varphi(0, 2^{j-1} x) + \varphi(2^{j-1} x, 2^{j-1} x)}{2^j} \\ &= \lim_{n\to\infty} \{(1 - \frac{1}{2^n})(\|g(0)\| + \|h(0)\|) \\ &+ \sum_{j=1}^{n} \frac{\varphi(2^{j-1} x, 0) + \varphi(0, 2^{j-1} x) + \varphi(2^{j-1} x, 2^{j-1} x)}{2^j} \} \\ &= \|g(0)\| + \|h(0)\| + \varepsilon(x) \ \ for \ \ all \ \ x \in G. \end{aligned} \tag{1.40}$$

Finally, we show that T is unique. If not, there exists $U : G \to X$ another such mapping with

$$U(x+y) = U(x) + U(y)$$

and (1.17) is satisfied. Then,

$$\begin{aligned} \|T(x) - U(x)\| &= \|2^{-n} T(2^n x) - 2^{-n} U(2^n x)\| \\ &\le \|2^{-n} T(2^n x) - 2^{-n} f(2^n x)\| + \|2^{-n} f(2^n x) - 2^{-n} U(2^n x)\| \\ &\le 2^{-n} (\|g(0)\| + \|h(0)\| + \varepsilon(2^n x)) + 2^{-n} (\|g(0)\| \\ &\quad + \|h(0)\| + \varepsilon(2^n x)) \end{aligned}$$

$$= \frac{\|g(0)\| + \|h(0)\|}{2^{n-1}}$$

$$+2\sum_{j=1}^{\infty} \frac{\varphi(2^{n+j-1}x, 0) + \varphi(0, 2^{n+j-1}x) + \varphi(2^{n+j-1}x, 2^{n+j-1}x)}{2^{n+j}}$$

$$= \frac{\|g(0)\| + \|h(0)\|}{2^{n-1}}$$

$$+ 2 \sum_{k=n+1}^{\infty} \frac{\varphi(2^{k-1}x, 0) + \varphi(0, 2^{k-1}x) + \varphi(2^{k-1}x, 2^{k-1}x)}{2^k} \tag{1.41}$$

for all $x \in G$. Taking the limit in (1.41) as $n \to \infty$, we have

$$T(x) = U(x)$$

due to (1.35) for all $x \in G$. This completes the proof of the theorem.

Corollary 1 *Let $f, g, h : G \to X$ be such that*

$$g(0) = 0, \;\; h(0) = 0, \;\; \|f(x+y) - g(x) - h(y)\| \leq \varphi(x, y)$$

for all x, $y \in G$. Then there exists a unique mapping $T : G \to X$ such that

$$T(x+y) = T(x) + T(y),$$
$$\|f(x) - T(x)\| \leq \varepsilon(x),$$
$$\|g(x) - T(x)\| \leq \varphi(x, 0) + \varepsilon(x)$$

and

$$\|h(x) - T(x)\| \leq \varphi(0, x) + \varepsilon(x) \;\; for \;\; all \;\; x, y \in G.$$

Proof 4 *Given that*

$$\|f(x+y) - g(x) - h(y)\| \leq \varphi(x, y).$$

Putting $x = 0$ and $y = 0$, we obtain

$$\|f(y) - h(y)\| \leq \varphi(0, y),$$

and

$$\|f(x) - g(x)\| \leq \varphi(x, 0)$$

respectively. Clearly,

$$\|f(2^n y)/2^n - h(2^n y)/2^n\| \leq \varphi(0, 2^n y)/2^n.$$

Taking limit as $n \to \infty$ on both sides, we find

$$\lim_{n\to\infty} \|f(2^n y)/2^n - h(2^n y)/2^n\| \leq \lim_{n\to\infty} \varphi(0, 2^n y)/2^n = 0,$$

that is,

$$\lim_{n\to\infty}\frac{f(2^n y)}{2^n}=\lim_{n\to\infty}\frac{h(2^n y)}{2^n}$$

due to (1.35). By Theorem 5, the sequence $\{f(2^n x)/2^n\}$ *is a Cauchy sequence. Let*

$$\lim_{n\to\infty}\frac{f(2^n x)}{2^n}=T(x).$$

Using (1.36) and (1.39), it follows that

$$T(x+y)=T(x)+T(y) \text{ for all } x,y\in G.$$

Hence T *is additive. As*

$$\|f(x)-T(x)\|\leq\|g(0)\|+\|h(0)\|+\varepsilon(x),$$

then $g(0)=h(0)=0$ *implies that*

$$\|f(x)-T(x)\|\leq\varepsilon(x),\ \forall\ x\in G.$$

Similarly,

$$\|g(x)-T(x)\|\leq\varphi(x,0)+\varepsilon(x),\ \forall\ x\in G$$

and

$$\|h(x)-T(x)\|\leq\varphi(0,x)+\varepsilon(x),\ \forall\ x\in G.$$

T *is unique due to (1.41). This complete the proof of the Corollary.*

Corollary 2 *Let* $\delta>0$*, and let* $f,g,h:G\to X$ *be such that*

$$g(0)=0,\ \ h(0)=0,\ \ \|f(x+y)-g(x)-h(y)\|\leq\delta$$

for all $x,\ y\in G$*. Then there exists a unique mapping* $T:G\to X$ *such that*

$$\begin{aligned}&T(x+y)=T(x)+T(y),\\&\|f(x)-T(x)\|\leq 3\delta,\\&\|g(x)-T(x)\|\leq 4\delta,\end{aligned}$$

and

$$\|h(x)-T(x)\|\leq 4\delta \text{ for all } x,y\in G.$$

Proof 5 *Since* $\sum_{j=1}^{\infty}2^{-j}\varphi(2^j x,2^j y)$ *is convergent, then we can find* $\delta>0$ *such that*

$$\sum_{j=1}^{\infty}2^{-j}\varphi(2^j x,2^j y)\leq\delta.$$

Given that

$$\|f(x+y)-g(x)-h(y)\|\leq\delta.$$

Putting $x = 0$ and $y = 0$, we obtain

$$\|f(y) - h(y)\| \leq \delta$$

and

$$\|f(x) - g(x)\| \leq \delta$$

respectively. Clearly,

$$\|f(2^n y)/2^n - h(2^n y)/2^n\| \leq \delta/2^n.$$

Taking limit as $n \to \infty$ on both sides, we find

$$\lim_{n\to\infty} \|f(2^n y)/2^n - h(2^n y)/2^n\| \leq \lim_{n\to\infty} \delta/2^n = 0,$$

that is,

$$\lim_{n\to\infty} f(2^n y)/2^n = \lim_{n\to\infty} h(2^n y)/2^n$$

due to (1.35). By Theorem 5, the sequence $\{f(2^n x)/2^n\}$ is a Cauchy sequence. Let

$$\lim_{n\to\infty} \frac{f(2^n x)}{2^n} = T(x).$$

Using (1.36) and (1.39), it follows that

$$T(x+y) = T(x) + T(y) \ \ \textit{for all} \ \ x, y \in G.$$

Hence T is additive. Indeed,

$$\begin{aligned}
\|f(x) - T(x)\| &\leq \varepsilon(x) \\
&= \sum_{j=1}^{\infty} 2^{-j}\varphi(2^j x, 2^j y) + \sum_{j=1}^{\infty} 2^{-j}\varphi(2^j x, 0) + \sum_{j=1}^{\infty} 2^{-j}\varphi(0, 2^j y) \\
&= 3\delta.
\end{aligned}$$

Similarly,

$$\|g(x) - T(x)\| \leq \varphi(x, 0) + \varepsilon(x) \ \ \textit{for all} \ \ x \in G$$

and

$$\|h(x) - T(x)\| \leq \varphi(0, x) + \varepsilon(x), \ \ \forall \ \ x \in G.$$

The proof of T is unique is similar as proved in Theorem 5.

Corollary 3 *Let V be a real or complex vector space and X a Banach space. And let $F, G, H : V \to X$ be mappings satisfying the inequality*

$$\|2F(x + y/2) - G(x) - H(y)\| \leq \varphi(x, y) \tag{1.42}$$

for all $x, y \in V$. Then there exists a unique additive mapping $T : V \to X$ such that

$$\begin{aligned}
\|2F(x/2) - T(x)\| &\leq \|G(0)\| + \|H(0)\| + \varepsilon(x), \\
\|G(x) - T(x)\| &\leq \|G(0)\| + 2\|H(0)\| + \varphi(x, 0) + \varepsilon(x)
\end{aligned}$$

and

$$\|H(x) - T(x)\| \leq 2\|G(0)\| + \|H(0)\|\varphi(0,x) + \varepsilon(x) \ \ for \ \ all \ \ x, y \in V.$$

Proof 6 *Putting $y = 0$ and $x = 0$ in(1.42), we obtain*

$$\|2F(x/2) - G(x) - H(0)\| \leq \varphi(x,0),$$

$$\|2F(y/2) - G(0) - H(y)\| \leq \varphi(0,y).$$

Let $2F(x/2) = f(x)$. Then $2F(x + y/2) = f(x + y)$ and hence

$$\|f(x) - G(x) - H(0)\| \leq \varphi(x,0).$$

Consequently,

$$\|f(2^n x)/2^n - G(2^n x)/2^n\| \leq [\|H(0)\| + \varphi(2^n x, 0)]/2^n.$$

Taking limit as $n \to \infty$ to the above relation, it follows that

$$\lim_{n\to\infty} f(2^n x)/2^n = \lim_{n\to\infty} G(2^n x)/2^n$$

and similarly,

$$\lim_{n\to\infty} f(2^n y)/2^n = \lim_{n\to\infty} H(2^n y)/2^n \ \ for \ \ all \ \ x, y \in V$$

due to (1.35). Using the fact that $\lim_{n\to\infty} f(2^n x/2^n) = T(x)$, it happens that

$$\begin{aligned} 0 = &\lim_{n\to\infty} \|f(2^n x + 2^n y)/2^n - G(2^n x)/2^n - H(2^n y)/2^n\| \\ &= T(x+y) - T(x) - T(y). \end{aligned}$$

Following (1.40), it concludes that

$$\|2F(x/2) - T(x)\| \leq \|G(0)\| + \|H(0)\| + \varepsilon(x).$$

Similarly,

$$\|G(x) - T(x)\| \leq \|G(0)\| + 2\|H(0)\| + \varphi(x,0) + \varepsilon(x)$$

and

$$\|H(x) - T(x)\| \leq 2\|G(0)\| + \|H(0)\| + \varphi(0,x) + \varepsilon(x)$$

because of (1.18) and (1.19), respectively. Uniqueness of T is similar as proved in Theorem 5. This complete the proof of the Corollary.

Corollary 4 *Let V be a real or complex vector space and X a Banach space. Let $a \neq 0$, $b \neq 0$ be two real or complex numbers. And let $F, G, H : V \to X$ be mappings satisfying the inequality*

$$\|F(ax + by) - aF(x) - bF(y)\| \leq \varphi(x,y)$$

for all $x, y \in V$. *Then there exists a unique additive mapping* $T : V \to X$ *such that*

$$\|F(x) - T(x)\| \leq (|a| + |b|)\|F(0)\| + \varepsilon(x),$$
$$\|aF(x/a) - T(x)\| \leq (|a| + 2|b|)\|F(0)\| + \varphi(x, 0) + \varepsilon(x),$$

and

$$\|bF(x/b) - T(x)\| \leq (2|a| + |b|)\|F(0)\| + \varphi(0, x) + \varepsilon(x)$$

for all $x, y \in V$.

Proof 7 *Given that*

$$\|F(ax + by) - aF(x) - bF(y)\| \leq \varphi(x, y)$$

for all $x, y \in V$. *Let* $ax = m, by = n$ *and* $aF(m/a) = g(m/a), bF(n/b) = h(n/b)$. *Then the last inequality becomes*

$$\|F(m + n) - g(m/a) - h(n/b)\| \leq \varphi(m/a, n/b).$$

When $m = 0$, *we obtain*

$$\|F(n) - g(0) - h(n/b)\| \leq \varphi(0, n/b)$$

and when $n = 0$, *we obtain*

$$\|F(m) - g(m/a) - h(0)\| \leq \varphi(m/a, 0).$$

Proceeding as in Theorem 5, it follows that

$$\|F(m/a) - T(m/a)\| \leq \|g(0)\| + \|h(0)\| + \varepsilon(x/a).$$

Hence,

$$\|F(x) - T(x)\| \leq |a|\|F(0)\| + |b|\|F(0)\| + \varepsilon(x).$$

Moreover,

$$\|aF(x/a) - T(x)\| \leq (|a| + 2|b|)\|F(0)\| + \varphi(x, 0) + \varepsilon(x),$$

and

$$\|bF(x/b) - T(x)\| \leq (2|a| + |b|)\|F(0)\| + \varphi(0, x) + \varepsilon(x) \quad for\ all\ x, y \in V.$$

Corollary 5 *Consider* E_1, E_2 *to be two Banach spaces, and let* $f, g, h : E_1 \to E_2$ *be mappings. Assume that there exists* $\theta \geq 0$ *and* $p \in [0, 1)$ *such that*

$$\|f(x + y) - g(x) - h(y)\| \leq \theta(\|x\|^p + \|y\|^p)$$

for all $x, y \in E_1$*.Then there exists a unique linear mapping* $T : E_1 \to E_2$ *such that*

$$\|f(x) - T(x)\| \leq \|g(0)\| + \|h(0)\| + \frac{4\theta}{2 - 2^p}\|x\|^p,$$
$$\|g(x) - T(x)\| \leq \|g(0)\| + 2\|h(0)\| + \frac{6 - 2^p}{2 - 2^p}\|x\|^p,$$
$$\|h(x) - T(x)\| \leq 2\|g(0)\| + \|h(0)\| + \frac{6 - 2^p}{2 - 2^p}\|x\|^p$$

for all $x \in E_1$*.*

Corollary 6 *Let* $0 \leq p + q < 1$*, where* p *and* q *are the non-negative real numbers, and let* $f, g, h : G \to X$ *be such that*

$$g(0) = 0, \quad h(0) = 0, \quad \|f(x + y) - g(x) - h(y)\| \leq \theta\|x\|^p\|y\|^q$$

for all $x, y \in G$*.Then there exists a unique mapping* $T : G \to X$ *such that*

$$T(x + y) = T(x) + T(y),$$
$$\|f(x) - T(x)\| \leq \frac{\theta}{2 - 2^{p+q}}\|x\|^{p+q},$$
$$\|g(x) - T(x)\| \leq \frac{\theta}{2 - 2^{p+q}}\|x\|^{p+q},$$

$$\|h(x) - T(x)\| \leq \frac{\theta}{2 - 2^{p+q}}\|x\|^{p+q}$$

for all $x, y \in G$*.*

1.1.4 Stability of $f(xy) + f(x + y) = f(xy + x) + f(y)$

In this section, we denote by F a ring with the unit element 1 and by E a Banach space. Let $\varphi : F \times F \to [0, \infty)$ be such that

$$\sum_{n=1}^{\infty} 2^{-n}\varphi(2^{n-1}x, 2^{n-1}y + z) < \infty, \quad \text{for all } x, y, z \in F. \tag{1.43}$$

Hyers-Ulam-Rassias stability of the Davison functional equation $f(xy) + f(x + y) = f(xy + x) + f(y)$ for a class of functions from a ring to a Banach space and also, we investigate the Davison equation of Pexider type.

Theorem 6 *If a function* $f : F \to E$ *satisfies the inequality*

$$\|f(xy) + f(x + y) - f(xy + x) - f(y)\| \leq \varphi(x, y) \tag{1.44}$$

for all $x, y \in F$, then there exists a unique additive function $A : F \to E$ such that

$$\|f(6x) - A(x) - f(0)\| \leq \sum_{n=0}^{\infty} \frac{M(2^n x)}{2^n}, \quad \forall \ x \in F, \tag{1.45}$$

where

$$\begin{aligned} M(x) = \frac{1}{2}[&\varphi(4x, -4x) + \varphi(4x, -4x+1) + \varphi(8x, -2x) + \varphi(3x, 0) \\ &+ \varphi(3x, 1) + \varphi(6x, 0) + \varphi(7x, -x) + \varphi(7x, -4x+1) + \varphi(14x, -2x)]. \end{aligned}$$

Proof 8 *If we replace y by $y+1$ in (1.44), we get*

$$\|f(xy+x) + f(x+y+1) - f(xy+2x) - f(y+1)\| \leq \varphi(x, y+1) \tag{1.46}$$

for any $x, y \in F$. Hence,

$$\begin{aligned} &\|f(xy) + f(x+y) + f(x+y+1) - f(y) - f(xy+2x) - f(y+1)\| \\ &\leq \|f(xy) + f(x+y) - f(xy+x) - f(y)\| \\ &+\|f(xy+x) + f(x+y+1) - f(xy+2x) - f(y+1)\| \\ &\leq \varphi(x, y) + \varphi(x, y+1), \ \textit{for all } x, y \in F \end{aligned}$$

due to (1.44) and (1.46). Replacing y by $4y$ in the last inequality, we obtain

$$\begin{aligned} &\|f(4xy) + f(x+4y) + f(x+4y+1) - f(4y) - f(4xy+2x) - f(4y+1)\| \\ &\leq \varphi(x, 4y) + \varphi(x, 4y+1), \ \ \textit{for all} \ \ x, y \in F. \end{aligned}$$

Therefore,

$$\begin{aligned} &\|f(x+4y) + f(x+4y+1) - f(2x+2y) - f(4y) - f(4y+1) + f(2y)\| \\ &\leq \|f(4xy) + f(x+4y) + f(x+4y+1) - f(4y) - f(4xy+2x) \\ &-f(4y+1)\| + \|f(4xy) + f(2x+2y) - f(4xy+2x) - f(2y)\| \\ &\leq \varphi(x, 4y) + \varphi(x, 4y+1) + \varphi(2x, 2y) \ \ \textit{for all} \ \ x, y \in F \end{aligned}$$

because of (1.44). If we replace x by $x-y$ in the last inequality, we find

$$\|f(x+3y) + f(x+3y+1) - f(2x) - f(4y) - f(4y+1) + f(2y)\|$$

$$\leq \varphi(x-y, 4y) + \varphi(x-y, 4y+1) + \varphi(2x-2y, 2y)$$

for all $x, y \in F$. Consequently,

$$\begin{aligned} &\|f(3x+3y) + f(3x+3y+1) - f(6x) - f(4y) \\ &\qquad -f(4y+1) + f(2y)\| \\ &\leq \varphi(3x-y, 4y) + \varphi(3x-y, 4y+1) + \varphi(6x-2y, 2y) \end{aligned} \tag{1.47}$$

for any $x, y \in F$ and substituting $3x$ for x.If we replace y by $-x$ in (1.47), it yields that

$$\|f(0)+f(1)-f(6x)-f(-4x)-f(-4x+1)+f(-2x)\| \leq \varphi(4x,-4x)+\varphi(4x,-4x+1)+\varphi(8x,-2x) \tag{1.48}$$

for every $x \in F$. If we replace y by 0 in (1.47), we have

$$\|f(3x)+f(3x+1)-f(6x)-f(1)\| \leq \varphi(3x,0)+\varphi(3x,1)+\varphi(6x,0) \tag{1.49}$$

for every $x \in F$. If we replace x, y by $2x, -x$ respectively in (1.47), we have

$$\|f(3x)+f(3x+1)-f(12x)-f(-4x)-f(-4x+1)+f(-2x)\| \leq \varphi(7x,-4x)+\varphi(7x,-4x+1)+\varphi(14x,-2x) \tag{1.50}$$

for every $x \in F$. From (1.48),(1.49) and (1.50) it follows that

$$\begin{aligned}
&\|\frac{f(12x)-f(0)}{2}-(f(6x)-f(0))\| \\
&= \frac{1}{2}\|2f(6x)-f(12x)-f(0)\| \\
&\leq \frac{1}{2}\Big[\|f(0)+f(1)-f(6x)-f(-4x)-f(-4x+1)+f(-2x)\| \\
&+ \|f(3x)+f(3x+1)-f(6x)-f(1)\| \\
&+ \|f(3x)+f(3x+1)-f(12x)-f(-4x)-f(-4x+1)+f(-2x)\|\Big] \\
&\leq \frac{1}{2}\Big[\varphi(4x,-4x)+\varphi(4x,-4x+1)+\varphi(8x,-2x)+\varphi(3x,0) \\
&+\varphi(3x,1)+\varphi(6x,0)+\varphi(7x,-x)+\varphi(7x,-4x+1)+\varphi(14x,-2x)\Big] \\
&= M(x) \ \ for \ \ every \ \ x \in F
\end{aligned} \tag{1.51}$$

implies that

$$\|\frac{f(2^n.6x)-f(0)}{2^n}-\frac{f(2^{n-1}.6x)-f(0)}{2^{n-1}}\| \leq \frac{M(2^{n-1}x)}{2^{n-1}}$$

for all $n \in N$ and $x \in F$. Replacing x by $2x$ in (1.51), we find

$$\|f(2.6x)-f(0)-\frac{f(2^2.6x)-f(0)}{2}\| \leq M(2x).$$

Hence,

$$\|[f(2.6x)-f(0)]/2-\frac{f(2^2.6x)-f(0)}{2^2}\| \leq M(2x)/2$$

implies that

$$\|f(6x) - f(0) - \frac{f(2^2.6x) - f(0)}{2^2}\| - \|f(6x) - f(0) - \frac{f(2.6x) - f(0)}{2}\| \leq \frac{M(2x)}{2}.$$

Consequently,

$$\|f(6x) - f(0) - \frac{f(2^2.6x) - f(0)}{2^2}\| \leq \frac{M(2x)}{2} + M(x) = \sum_{i=1}^{2} \frac{M(2^{i-1}x)}{2^{i-1}}.$$

Further,

$$\begin{aligned}\|f(6x) - f(0) - \frac{f(2^3.6x) - f(0)}{2^3}\| &\leq \|f(6x) - f(0) - \frac{f(2^2.6x) - f(0)}{2^2}\| \\ &+ \|\frac{f(2^2.6x) - f(0)}{2^2} - \frac{f(2^3.6x) - f(0)}{2^3}\| \\ &= \sum_{i=1}^{2} \frac{M(2^{i-1}x)}{2^{i-1}} + \frac{M(2^2x)}{2^2} = \sum_{i=1}^{3} \frac{M(2^{i-1}x)}{2^{i-1}}\end{aligned}$$

implies that

$$\|f(6x) - f(0) - \frac{f(2^n.6x) - f(0)}{2^n}\| \leq \sum_{i=1}^{n} \frac{M(2^{i-1}x)}{2^{i-1}} \tag{1.52}$$

inductively, for all $n \in N$ *and* $x \in F$. *Now,*

$$\begin{aligned}&\|\frac{f(2^n.6x) - f(0)}{2^n} - \frac{f(2^{m+n}.6x) - f(0)}{2^{m+n}}\| \\ &\leq \|\frac{f(2^n.6x) - f(0)}{2^n} - \frac{f(2^{n+1}.6x) - f(0)}{2^{n+1}}\| \\ &+ \|\frac{f(2^{n+1}.6x) - f(0)}{2^{n+1}} - \frac{f(2^{n+2}.6x) - f(0)}{2^{n+2}}\| \\ &+ \cdots + \|\frac{f(2^{m+n-1}.6x) - f(0)}{2^{m+n-1}} - \frac{f(2^{m+n}.6x) - f(0)}{2^{m+n}}\| \\ &= \frac{M(2^nx)}{2^n} + \cdots + \frac{M(2^{m+n}x)}{2^{m+n}} \\ &= \sum_{i=n+1}^{m+n} \frac{M(2^{i-1}x)}{2^{i-1}} < \infty\end{aligned}$$

due to (1.43) and (1.51) implies that $\{\frac{f(2^n6x) - f(0)}{2^n}\}$ *is a Cauchy sequence for* $x \in F$. *If* $A : F \to E$, *then we let*

$$\lim_{n\to\infty} \frac{f(2^n6x) - f(0)}{2^n} = A(x)$$

for $x \in F$ and we get (1.45) from (1.52), ultimately. If we replace x, y by $x+y, -x$ respectively in (1.47), we obtain

$$\begin{aligned}&\|f(3y) + f(3y+1) - f(6x+6y) - f(-4x) - f(-4x+1) + f(-2x)\| \\ &\leq \varphi(4x+3y, -4x) + \varphi(4x+3y, -4x+1) + \varphi(8x+6y, -2x) \qquad (1.53)\end{aligned}$$

for every $x, y \in F$, and if we replace x, y by $x+y, -y$ respectively in (1.47), we get

$$\begin{aligned}&\|f(3x) + f(3x+1) - f(6x+6y) - f(-4y) - f(-4y+1) + f(-2y)\| \\ &\leq \varphi(3x+4y, -4y) + \varphi(3x+4y, -4y+1) + \varphi(6x+8y, -2y) \qquad (1.54)\end{aligned}$$

for every $x, y \in F$. Furthermore, if we replace x, y by $2y, -y$ respectively in (1.47),we get

$$\begin{aligned}&\|f(3y) + f(3y+1) - f(12y) - f(-4y) - f(-4y+1) + f(-2y)\| \\ &\leq \varphi(7y, -4y) + \varphi(7y, -4y+1) + \varphi(14y, -2y) \qquad (1.55)\end{aligned}$$

for every $y \in F$. Lastly, if we replace x, y by $2x, -x$, respectively in (1.47), we observe that

$$\begin{aligned}&\|f(3x) + f(3x+1) - f(12x) - f(-4x) - f(-4x+1) + f(-2x)\| \\ &\leq \varphi(7x, -4x) + \varphi(7x, -4x+1) + \varphi(14x, -2x) \qquad (1.56)\end{aligned}$$

for every $x \in F$. Using (1.53), (1.54), (1.55) and (1.56), we find

$$\begin{aligned}&\|2f(6x+6y) - f(12x) - f(12y)\| \\ &\leq \|f(6x+6y) - f(3y) - f(3y+1) + f(-4x) + f(-4x+1) - f(-2x)\| \\ &+ \|f(3y) + f(3y+1) - f(12y) - f(-4y) - f(-4y+1) + f(-2y)\| \\ &+ \|f(6x+6y) - f(3x) - f(3x+1) + f(-4y) + f(4y+1) - f(-2y)\| \\ &+ \|f(3x) + f(3x+1) - f(12x) - f(-4x) - f(-4x+1) + f(-2x)\| \\ &= \|f(3y) + f(3y+1) - f(6x+6y) - f(-4x) - f(-4x+1) + f(-2x)\| \\ &+ \|f(3x) + f(3x+1) - f(6x+6y) - f(-4y) - f(4y+1) + f(-2y)\| \\ &+ \|f(3y) + f(3y+1) - f(12y) - f(-4y) - f(-4y+1) + f(-2y)\| \\ &+ \|f(3x) + f(3x+1) - f(12x) - f(-4x) - f(-4x+1) + f(-2x)\| \\ &\leq \varphi(4x+3y, -4x) + \varphi(4x+3y, -4x+11) + \varphi(8x+6y, -2x) \\ &+ \varphi(3x+4y, -4y) + \varphi(3x+4y, -4y+1) + \varphi(6x+8y, -2y) \\ &+ \varphi(7y, -4y) + \varphi(7y, -4y+1) + \varphi(14y, -2x) =: M'(x, y) \qquad (1.57)\end{aligned}$$

for every $x, y \in F$. Replacing x, y by $2^n x, 2^n y$ in (1.57) and then dividing the resulting inequality by 2^n, we find

$$\left\|\frac{f(2^n.6x + 2^n.6y)}{2^{n-1}} - \frac{f(2^n.12x)}{2^n} - \frac{f(2^n 12y)}{2^n}\right\| \leq \frac{M'(2^n x, 2^n y)}{2^n},$$

for every $x, y \in F$. From the given hypothesis and the definition of A, it follows that

$$2A(x+y) - A(2x) - A(2y) = 0$$

for every $x, y \in F$. Since $A(2x) = 2A(x)$, then $A(x+y) = A(x) + A(y)$ for every $x, y \in F$. We claim that A is unique. Let $B : F \to E$ be another additive mapping satisfying (1.45). Then

$$\begin{aligned}\|A(x) - B(x)\| &\leq \|\frac{f(6.2^n x) - f(0)}{2^n} - A(x)\| + \|\frac{f(6.2^n x) - f(0)}{2^n} - B(x)\| \\ &= \|\frac{f(6.2^n x) - f(0) - A(2^n x)}{2^n}\| + \|\frac{f(6.2^n x) - f(0) - B(2^n x)}{2^n}\| \\ &\leq \frac{1}{2^n}\sum_{i=0}^{\infty}\frac{M(2^i x)}{2^{i-1}}, \ [\because A \text{ and } B \text{ are additive.}]\end{aligned}$$

for every $x \in F$. Taking the limit as $n \to \infty$ in the last inequality, we obtain that $A(x) = B(x)$, for all $x \in F$. This completes the proof of the theorem.

Corollary 7 *If a function $f : F \to E$ satisfies the inequality*

$$\|f(xy) + f(x+y) - f(xy+x) - f(y)\| \leq \varepsilon$$

for all $x,y \in F$, then there exists a unique additive function $A : F \to E$ such that

$$\|f(6x) - A(x) - f(0)\| \leq 9\varepsilon$$

for all $x \in F$.

Proof 9 *Since $\sum_{n=1}^{\infty} 2^{-n}\varphi(2^{n-1}x, 2^{n-1}y + z) < \infty$, then there exists $\varepsilon > 0$ such that*

$$\sum_{n=1}^{\infty} 2^{-n}\varphi(2^{n-1}x, 2^{n-1}y + z) \leq \varepsilon.$$

Hence,

$$\|f(xy) + f(x+y) - f(xy+x) - f(y)\| \leq \varepsilon.$$

From Theorem 6, we know that

$$\|f(6x) - A(x) - f(0)\| \leq \sum_{n=0}^{\infty}\frac{M(2^n x)}{2^n}$$

$$\begin{aligned}= \sum_{n=0}^{\infty}\frac{1}{2^{n+1}}\Big[&\varphi(2^n.4x, -2^n.4x) + \varphi(2^n.4x, -2^n.4x+1) + \varphi(2^n.8x, -2^n.2x) \\ &+ \varphi(3.2^n x, 0) + \varphi(2^n.3x, 1) + \varphi(2^n.6x, 0) \\ &+ \varphi(2^n.7x, -2^n.x) + \varphi(2^n.7x, -2^n.4x+1) + \varphi(2^n.14x, -2^n.2x)\Big] \\ &= 9\varepsilon \ \text{ for all } \ x \in F.\end{aligned}$$

This completes the proof of the Corollary.

Corollary 8 *Let F be a normed algebra and let $0 \le p < 1$. If a function $f : F \to E$ satisfies the inequality*

$$\|f(xy) + f(x+y) - f(xy+x) - f(y)\| \le \theta(\|x\|^p + \|y\|^p)$$

for all $x, y \in F$, then there exists a unique additive function $A : F \to E$ such that

$$\begin{aligned}\|f(6x) - A(x) - f(0)\| \le{}& \tfrac{1}{2-2^p}\theta[(2.2^p + 2.3^p + 5.4^p + 6^p \\ &+2.7^p + 8^p + 14^p + 1)\|x\|^p] + 3.\|1\|^p\end{aligned}$$

for all $x \in F$.

Now we prove the stability of Davison functional equation of Pexider type. Let $\varphi : F \times F \to [0, \infty)$ be a function and let $\psi : F \times F \to [0, \infty)$ be a function defined by

$$\psi(x, y) = \varphi(x, y) + \varphi(xy + x, 0) + \varphi(0, y) + \varphi(1, xy) + \varphi(xy + 1, 0) + \varphi(0, xy)$$

for all $x, y \in F$.

Theorem 7 *If the function $f, g, h, k : F \to E$ satisfy the inequality*

$$\|f(xy) + g(x+y) - h(xy+x) - k(y)\| \le \varphi(x, y) \tag{1.58}$$

for all $x, y \in F$ and ψ is a function such that

$$\sum_{n=1}^{\infty} 2^{-n}\psi(2^{n-1}x, 2^{n-1}y + z) < \infty$$

for all $x, y, z \in F$, then there exists a unique additive function $A : F \to E$ such that

$$\|f(6x) - A(x) - f(0)\| \le \sum_{n=0}^{\infty} \frac{M(2^n x)}{2^n} + \varphi(1, 6x) \tag{1.59}$$

$$+ \varphi(6x + 1, 0) + \varphi(0, 6x) + \varphi(0, 0)$$

$$\|g(6x) - A(x) - g(0)\| \le \sum_{n=0}^{\infty} \frac{M(2^n x)}{2^n} \tag{1.60}$$

$$\|h(6x) - A(x) - h(0)\| \le \sum_{n=0}^{\infty} \frac{M(2^n x)}{2^n} + \varphi(0, 0) + \varphi(6x, 0) \tag{1.61}$$

$$\|k(6x) - A(x) - k(0)\| \le \sum_{n=0}^{\infty} \frac{M(2^n x)}{2^n} + \varphi(0, 0) + \varphi(0, 6x) \tag{1.62}$$

for all $x \in F$, where

$$
\begin{aligned}
M(x) = \frac{1}{2}[&4\varphi(0,0) + \varphi(0,1) + \varphi(0,-x) + \varphi(0,-2x) \\
&+ \varphi(0,-2x) + \varphi(0,3x) + \varphi(0,-4x) + 2\varphi(0,-4x+1) \\
&+ \varphi(0,-7x^2) + 2\varphi(0,-16x^2) + \varphi(0,-16x^2+4x) + \varphi(0,-28x^2) \\
&+ \varphi(0,-28x^2+7x) + 4\varphi(1,0) + \varphi(1,3x) + \varphi(1,-7x^2) \\
&+ 2\varphi(1,-16x^2) + \varphi(1,-16x^2+4x) + \varphi(1,-28x^2) \\
&+ \varphi(1,-28x^2+7x) + 2\varphi(3x,0) + \varphi(3x,1) + \varphi(3x+1,0) \\
&+ \varphi(4x,-4x) + \varphi(4x,-4x+1) + 3\varphi(6x,0) + \varphi(7x,-x) \\
&+ \varphi(7x,-4x+1) + \varphi(8x,-2x) + \varphi(14x,-2x) \\
&+ \varphi(-7x^2+1,0) + \varphi(-7x^2+7x,0) + 2\varphi(-16x^2+4x+1,0) \\
&+ \varphi(-16x^2+4x,0) + \varphi(-16x^2+4x+1,0) \\
&+ 2\varphi(-16x^2+8x,0) + \varphi(-28x^2+1,0) \\
&+ \varphi(-28x^2+7x+1,0) + 2\varphi(-28x^2+14x,0)].
\end{aligned}
$$

Proof 10 *Assume that the functions $f, g, h, k : F \to E$ satisfy (1.58). Putting $y = 0, x = 0$ and $x = 1$ separately in (1.58), we find*

$$
\begin{aligned}
&\|f(0) + g(x) - h(x) - k(0)\| \leq \varphi(x,0), \\
&\|f(0) + g(y) - h(0) - k(y)\| \leq \varphi(0,y) \qquad (1.63)
\end{aligned}
$$

and

$$
\|f(y) + g(y+1) - h(y+1) - k(y)\| \leq \varphi(1,y).
$$

Now,

$$
\begin{aligned}
\|f(y) - g(y) + h(0) + k(0) - 2f(0)\| &\leq \|f(y) + g(y+1) - h(y+1) - k(y)\| \\
&+ \| - g(y+1) + h(y+1) + k(0) - f(0)\| \\
&+ \| - g(y) + k(y) + h(0) - f(0)\| \\
&\leq \varphi(1,y) + \varphi(y+1,0) + \varphi(0,y) \qquad (1.64)
\end{aligned}
$$

due to (1.63). Hence,

$$
\begin{aligned}
&\|g(xy) + g(x+y) - g(xy+x) - g(y)\| \\
&\leq \|f(xy) + g(x+y) - h(xy+x) - k(y)\| \\
&+ \| - f(0) - g(xy+x) + h(xy+x) + k(0)\| \\
&+ \| - f(0) - g(y) + h(0) + k(y)\| \\
&+ \| - f(xy) + g(xy) - h(0) - k(0) + 2f(0)\| \\
&\leq \varphi(x,y) + \varphi(xy+x,0) + \varphi(0,y) + \varphi(1,xy) \\
&+ \varphi(xy+1,0) + \varphi(0,xy) \qquad (1.65)
\end{aligned}
$$

due to (1.58),(1.63) and (1.64). Furthermore, if we replace y *by* $y+1$ *in (1.65), we get*

$$\|g(xy+x)+g(x+y+1)-g(xy+2x)-g(y+1)\|$$
$$\le \varphi(x,y+1)+\varphi(xy+2x,0)+\varphi(0,y+1)$$
$$+\varphi(1,xy+x)+\varphi(xy+x+1,0)+\varphi(0,xy+x)$$
$$\|g(xy)+g(x+y)+g(x+y+1)-g(y)-g(xy+2x)-g(y+1)\|$$
$$\le \|g(xy)+g(x+y)-g(xy+x)-g(y)\|$$
$$+\|g(xy+x)+g(x+y+1)-g(xy+2x)-g(y+1)\|$$
$$\le \varphi(x,y)+\varphi(xy+x,0)+\varphi(0,y)+\varphi(1,xy)$$
$$+\varphi(xy+1,0)+\varphi(0,xy)+\varphi(x,y+1)+\varphi(xy+2x,0)$$
$$+\varphi(0,y+1)+\varphi(1,xy+x)+\varphi(xy+x+1,0)+\varphi(0,xy+x) \quad (1.66)$$

for all $x,y\in F$*. Lastly, if we replace* y *by* $4y$ *in (1.65), we get*

$$\|g(4xy)+g(x+4y)+g(x+4y+1)-g(4y)-g(4xy+2x)-g(4y+1)\|$$
$$\le \varphi(x,4y)+\varphi(4xy+x,0)+\varphi(0,4y)+\varphi(1,4xy)$$
$$+\varphi(4xy+1,0)+\varphi(0,4xy)+\varphi(x,4y+1)+\varphi(4xy+2x,0)$$
$$+\varphi(0,4y+1)+\varphi(1,4xy+x)+\varphi(4xy+x+1,0)+\varphi(0,4xy+x)$$

for all $x,y\in F$*. Using (1.65) and the last relation we find,*

$$\|g(x+4y)+g(x+4y+1)-g(2x+2y)-g(4y)-g(4y+1)+g(2y)\|$$
$$\le \|g(4xy)+g(x+4y)+g(x+4y+1)-g(4y)-g(4xy+2x)-g(4y+1)\|$$
$$+\|g(4xy)+g(2x+2y)-g(4xy+2x)-g(2y)\|$$
$$\le \varphi(x,4y)+\varphi(4xy+x,0)+\varphi(0,4y)+\varphi(1,4xy)$$
$$+\varphi(4xy+1,0)+\varphi(0,4xy)+\varphi(x,4y+1)+\varphi(4xy+2x,0)$$
$$+\varphi(0,4y+1)+\varphi(1,4xy+x)+\varphi(4xy+x+1,0)+\varphi(0,4xy+x)$$
$$+\varphi(2x,2y)+\varphi(4xy+2x,0)+\varphi(0,2y)+\varphi(1,4xy)$$
$$+\varphi(4xy+1,0)+\varphi(0,4xy).$$

Proceeding as in the proof of Theorem 6, we find

$$\|\frac{g(12x)-g(0)}{2}-(g(6x)-g(0))\|$$
$$=\frac{1}{2}\|2g(6x)-g(12x)-g(0)\|$$
$$\le \frac{1}{2}[\|g(0)+g(1)-g(6x)-g(-4x)-g(-4x+1)+g(-2x)\|$$
$$+\|g(3x)+g(3x+1)-g(6x)-g(1)\|$$
$$+\|g(3x)+g(3x+1)-g(12x)-g(-4x)-g(-4x+1)+g(-2x)\|]$$
$$\le \frac{1}{2}[4\varphi(0,0)+\varphi(0,1)+\varphi(0,-x)+\varphi(0,-2x)$$

$$
\begin{aligned}
&+\varphi(0,-2x)+\varphi(0,3x)+\varphi(0,-4x)+2\varphi(0,-4x+1)\\
&+\varphi(0,-7x^2)+2\varphi(0,-16x^2)+\varphi(0,-16x^2+4x)\\
&+\varphi(0,-28x^2)+\varphi(0,-28x^2+7x)\\
&+4\varphi(1,0)+\varphi(1,3x)+\varphi(1,-7x^2)+2\varphi(1,-16x^2)\\
&+\varphi(1,-16x^2+4x)+\varphi(1,-28x^2)+\varphi(1,-28x^2+7x)\\
&+2\varphi(3x,0)+\varphi(3x,1)+\varphi(3x+1,0)\\
&+\varphi(4x,-4x)+\varphi(4x,-4x+1)+3\varphi(6x,0)+\varphi(7x,-x)\\
&+\varphi(7x,-4x+1)+\varphi(8x,-2x)+\varphi(14x,-2x)+\varphi(-7x^2+1,0)\\
&+\varphi(-7x^2+7x,0)+2\varphi((-16x^2+1,0)+\varphi(-16x^2+4x,0)\\
&+\varphi(-16x^2+4x+1,0)+2\varphi(-16x^2+8x,0)+\varphi(-28x^2+1,0)\\
&+\varphi(-28x^2+7x+1,0)+2\varphi(-28x^2+14x,0)=M(x). \qquad (1.67)
\end{aligned}
$$

Using $2^{n-1}x$ for x in (1.67) and then dividing by 2^{n-1}, the resulting inequality becomes

$$\|\frac{g(2^n.6x)-g(0)}{2^n}-\frac{g(2^{n-1}.6x)-g(0)}{2^{n-1}}\|\leq\frac{M(2^{n-1}x)}{2^{n-1}},$$

$$\|g(6x)-g(0)-\frac{g(2^n.6x)-g(0)}{2^n}\|\leq\sum_{i=1}^{n}\frac{M(2^{i-1}x)}{2^{i-1}} \qquad (1.68)$$

and thus $\{\frac{g(2^n.6x)-g(0)}{2^n}\}$ is a Cauchy sequence for $x\in F$. If $A:F\to E$ defined by

$$A(x)=\lim_{n\to\infty}\frac{g(2^n.6x)-g(0)}{2^n},$$

then

$$\|g(6x)-A(x)-g(0)\|\leq\sum_{n=0}^{\infty}\frac{M(2^nx)}{2^n}.$$

From (1.60) and (1.63), we have

$$
\begin{aligned}
\|h(6x)-h(0)-A(x)\| &\leq \|h(6x)-g(6x)-f(0)+k(0)\|\\
&\quad+\|g(6x)-A(x)-g(0)\|+\|f(0)+g(0)-h(0)-k(0)\|\\
&\leq\sum_{n=0}^{\infty}\frac{M(2^nx)}{2^n}+\varphi(0,0)+\varphi(6x,0)
\end{aligned}
$$

and

$$
\begin{aligned}
\|k(6x)-A(x)-k(0)\| &\leq \|k(6x)-g(x)-f(0)+h(0)\|\\
&\quad+\|g(6x)-A(x)-g(0)\|+\|g(0)+f(0)-h(0)-k(0)\|\\
&\leq\sum_{n=0}^{\infty}\frac{M(2^nx)}{2^n}+\varphi(0,0)+\varphi(0,6x)
\end{aligned}
$$

for all $x\in F$. This completes the proof of the theorem.

Corollary 9 *If a function $f : F \to E$ satisfies the inequality*

$$\|f(xy) + g(x+y) - h(xy+x) - k(y)\| \leq \varepsilon$$

for all $x, y \in F$, then there exists a unique additive function $A : F \to E$ such that

$$\|f(6x) - A(x) - f(0)\| \leq 57\varepsilon$$

for all $x \in F$.

1.1.5 Stability of $f(x+iy) + f(x-iy) = 2f(x) - 2f(y)$

Let X, Y be complex vector spaces. It is shown that if a mapping $f : X \to Y$ satisfies

$$f(x+iy) + f(x-iy) = 2f(x) - 2f(y) \tag{1.69}$$

or

$$f(x+iy) - f(ix+y) = 2f(x) - 2f(y) \tag{1.70}$$

for all $x, y \in X$, then the mapping $f : X \to Y$ satisfies

$$f(x+y) + f(x-y) = 2f(x) + 2f(y) \ for \ all \ x, y \in X.$$

Again, we prove the generalized Hyers-Ulam stability of the functional equations (1.69) and (1.70) in complex Banach spaces.

In this section, we solve the functional equations (1.69) and (1.70) and prove the generalized Hyers-Ulam stability by using the fixed point method. Here, we assume that X and Y are complex vector spaces.

Proposition 1 *If a mapping $f : X \to Y$ satisfies*

$$f(x+iy) + f(x-iy) = 2f(x) - 2f(y) \ for \ all \ x, y \in X, \tag{1.71}$$

then the mapping $f : X \to Y$ is quadratic, that is,

$$f(x+y) + f(x-y) = 2f(x) + 2f(y)$$

holds for all $x \in X$.If a mapping $f : X \to Y$ is quadratic and $f(ix) = -f(x)$ holds for all $x \in X$, then the mapping $f : X \to Y$ satisfies (1.71).

Proof 11 *Suppose that $f : X \to Y$ satisfies (1.71). Putting $x = y$ in (1.69), we find*

$$f((1+i)x) + f((1-i)x) = 0 \ for \ all \ x \in X.$$

Replacing x by $(1+i)x$ in the above result, we find

$$f((1+i)(1+i)) + f((1-i)(1+i)x) = 0$$

implies that

$$f(2ix) + f(2x) = 0 \ for \ all \ x \in X.$$

Replacing x *by* $x/2$ *in the above result, we find*

$$f(ix) + f(x) = 0 \quad for \ all \ \ x \in X.$$

Then $f(ix) = -f(x)$. *Therefore,*

$$\begin{aligned} f(x+iy) + f(x-iy) &= 2f(x) - 2f(y) \\ &= 2f(x) + 2f(iy) \quad for \ all \ \ x, y \in X. \end{aligned} \tag{1.72}$$

Putting $z = iy$ *in (1.72), we get*

$$f(x+z) + f(x-z) = 2f(x) + 2f(z) \quad for \ all \ \ x, z \in X.$$

Conversely assume that a quadratic mapping $f : X \to Y$ *satisfies* $f(ix) = -f(x)$ *for all* $x \in X$. *Then*

$$f(x+iy) + f(x-iy) = 2f(x) + 2f(iy) = 2f(x) - 2f(y)$$

for all $x, y \in X$. *So the mapping* $f : X \to Y$ *satisfies (1.72).*

Proposition 2 *If a mapping* $f : X \to Y$ *satisfies* $f(0) = 0$ *and*

$$f(x+iy) - f(ix+y) = 2f(x) - 2f(y) \tag{1.73}$$

for all $x, y \in X$, *then the mapping* $f : X \to Y$ *is quadratic. If a mapping* $f : X \to Y$ *is quadratic and* $f(ix) = -f(x)$ *holds for all* $x \in X$, *then the mapping* $f : X \to Y$ *satisfies*

$$f(x+iy) - f(ix+y) = 2f(x) - 2f(y).$$

Proof 12 *Assuming that* $f : X \to Y$ *satisfies the equation* (1.73). *Putting* $y = 0$ *in (1.73), we find*

$$f(x) - f(ix) = 2f(x) \quad for \ all \ \ x \in X$$

and hence $f(ix) = -f(x)$ *for all* $x \in X$. *Therefore,*

$$\begin{aligned} f(x+iy) + f(x-iy) = f(x+iy) - f(ix+y) &= 2f(x) - 2f(y) \\ = 2f(x) - 2f(iy) \quad for \ all \ \ x, y \in X. \end{aligned} \tag{1.74}$$

Putting $z = iy$ *in (1.74), we get*

$$f(x+z) + f(x-z) = 2f(x) + 2f(z) \quad for \ all \ \ x, z \in X.$$

Suppose that a quadratic mapping $f : X \to Y$ *satisfies* $f(ix) = -f(x)$ *for all* $x \in X$. *Then*

$$\begin{aligned} f(x+iy) - f(ix+y) &= f(x+iy) + f(x-iy) \\ &= 2f(x) + 2f(iy) \\ &= 2f(x) - 2f(y) \end{aligned}$$

for all $x, y \in X$. *Hence, the mapping* $f : X \to Y$ *satisfies* $f(x+iy) - f(ix+y) = 2f(x) - 2f(y)$.

In the following we assume that X is a normed vector space and Y is a Banach space. For a given mapping $f : X \to Y$, we define

$$Cf(x,y) = f(x+iy) + f(x-iy) - 2f(x) + 2f(y), \ \forall \ x,y \in X.$$

Theorem 8 *Let $p < 2$, and θ be positive numbers, and $f : X \to Y$ be a mapping satisfying $f(ix) = -f(x)$ and*

$$\|Cf(x,y)\| \leq \theta(\|x\|^p + \|y\|^p) \tag{1.75}$$

for all $x, y \in X$. Then there exists a unique quadratic mapping $Q : X \to Y$ such that

$$\|f(x) - Q(x)\| \leq \frac{2\theta}{4-2^p}\|x\|^p \ \ for\ all \ \ x,y \in X. \tag{1.76}$$

Proof 13 *Since $f(ix) = -f(x)$ for all $x \in X$, then $f(0) = 0$. So, $f(-x) = f(i^2x) = -f(ix) = f(x)$ for all $x \in X$. Putting $y = -ix$ in (1.75), we find*

$$\|f(x+x) + f(x-x) - 2f(x) - 2f(-ix)\| \leq \theta(\|x\|^p + \|-ix\|^p),$$

that is,

$$\|f(2x) - 4f(x)\| \leq 2\theta\|x\|^p \tag{1.77}$$

for all $x \in X$. Hence,

$$\|f(x) - \frac{1}{4}f(2x)\| \leq \frac{\theta}{2}\|x\|^p$$

for all $x \in X$. So,

$$\begin{aligned}
\|\frac{1}{4^n}f(2^nx) - \frac{1}{4^m}f(2^mx)\| &= \|\frac{1}{4^n}f(2^nx) - \frac{1}{4^m}f(2^mx) + \frac{1}{4^{m+1}}f(2^{m+1}x) \\
&\quad - \frac{1}{4^{m+1}}f(2^{m+1}x) + \cdots - \frac{1}{4^{n-1}}f(2^{n-1}x) + \frac{1}{4^{n-1}}f(2^{n-1}x)\| \\
&\leq \|\frac{1}{4^m}f(2^mx) - \frac{1}{4^{m+1}}f(2^{m+1}x)\| + \cdots \\
&\quad + \|\frac{1}{4^{n-1}}f(2^{n-1}x) - \frac{1}{4^n}f(2^nx)\| \\
&= \sum_{j=m}^{n-1} \|\frac{1}{4^j}f(2^jx) - \frac{1}{4^{j+1}}f(2^{j+1}x)\| = \sum_{j=m}^{n-1}\frac{1}{4^j}\|f(2^jx) - \frac{1}{4}f(2^{j+1}x)\| \\
&\leq \sum_{j=m}^{n-1}\frac{1}{4^j}\cdot\frac{\theta}{2}\|2^jx\|^p = \sum_{j=m}^{n-1}\frac{2^{pj}\theta}{2^{2j+1}}\|x\|^p \\
&= \left(\frac{2^{pm}\theta}{2^{2m+1}} + \cdots + \frac{2^{p(n-1)}\theta}{2^{2(n-1)+1}}\right)\|x\|^p \\
&= \frac{2^{pm}\theta}{2^{2m+1}}\left(1 + 2^p/2^2 + \cdots + 2^{p(n-m-1)}/2^{2(n-m-1)}\right)\|x\|^p \\
&= \frac{2^{pm}4\theta\|x\|^p}{2^{2m+1}}\frac{1-(2^p/2^2)^{n-m}}{4-2^p}
\end{aligned} \tag{1.78}$$

for all nonnegative integers m and n with $n > m$. Taking limit as $n \to \infty$ and m as fixed in (1.78), the right side converges due to $p < 2$ and hence $\{\frac{1}{4^n} f(2^n x)\}$ is a Cauchy sequence for all $x \in X$. Let there be a mapping $Q : X \to Y$ such that

$$Q(x) = \lim_{n\to\infty} \frac{1}{4^n} f(2^n x), \ \ \forall \ x \in X.$$

From (1.75), it follows that

$$\begin{aligned}\|Cf(2^n x, 2^n y)\| \quad &\leq \theta(\|2^n x\| + \|2^n y\|^p) \\ &= \theta 2^{np}(\|x\|^p + \|y\|^p)\end{aligned}$$

and

$$\|\frac{1}{4^n} Cf(2^n x, 2^n y)\| \leq \frac{2^{pn}\theta}{4^n}(\|x\|^p + \|y\|^p),$$

$$\|CQ(x,y)\| = \lim_{n\to\infty} \frac{1}{4^n}\|Cf(2^n x, 2^n y)\|, \ \ \forall \ x, y \in X$$

implies that

$$\begin{aligned}\|CQ(x,y)\| &\leq \lim_{n\to\infty} \frac{2^{pn}\theta}{4^n}(\|x\|^p + \|y\|^p) \\ &= 0, \ \ \forall \ x, y \in X.\end{aligned}$$

So, $CQ(x,y) = 0$. By Proposition 1, the mapping $Q : X \to Y$ is quadratic. Putting $m = 0$ and taking limit as $n \to \infty$ in (1.78), we get

$$\lim_{n\to\infty} \|f(x) - \frac{1}{4^n} f(2^n x)\| \leq \lim_{n\to\infty} \sum_{j=0}^{n-1} \frac{2^{pj}\theta}{2^{2j+1}}\|x\|^p,$$

that is,

$$\begin{aligned}\|f(x) - Q(x)\| &\leq \sum_{j=0}^{\infty} \frac{2^{pj}\theta}{2^{2j+1}}\|x\|^p \\ &= \frac{\theta}{2}(1 + 2^p/2^2 + \cdots)\|x\|^p \\ &= \frac{4\theta\|x\|^p}{2(4-2^p)} = \frac{2\theta}{4-2^p}\|x\|^p.\end{aligned}$$

Next, to show that Q is unique. If not, there exists $T : X \to Y$ such that

$$T(x+iy) + T(x-iy) = 2T(x) - 2T(y)$$

and

$$\|f(x) - T(x)\| \leq \frac{2\theta}{4-2^p}\|x\|^p, \ \ \forall \ x, y \in X.$$

Since Q is quadratic, then by definition (Proposition 1) $4Q(x) = Q(2x)$ and hence by the induction method $4^n Q(x) = Q(2^n x)$, for all $x \in X$. Hence,

$$\begin{aligned}\|Q(x) - T(x)\| &= \frac{1}{4^n}\|Q(2^n x) - T(2^n x)\| \\ &\leq \frac{1}{4^n}(\|Q(2^n x) - f(2^n x)\| + \|T(2^n x) - f(2^n x)\| \\ &\leq \frac{1}{4^n}2.\frac{2\theta 2^{pn}}{4-2^p}\|x\|^p = \frac{4\theta}{4-2^p}\frac{2^{pn}}{4^n}\|x\|^p\end{aligned}$$

implies that $\lim_{n\to\infty}\|Q(x) - T(x)\| = 0$ *for all* $x \in X$*. This completes the proof of the theorem.*

Theorem 9 *Let $p > 2$, and θ be positive numbers, and $f : X \to Y$ be a mapping satisfying (1.75) and $f(ix) = -f(x)$ for all $x, y \in X$. Then there exists a unique quadratic mapping $Q : X \to Y$ such that*

$$\|f(x) - Q(x)\| \leq \frac{2\theta}{2^p - 4}\|x\|^p, \quad \forall\ x, y \in X. \tag{1.79}$$

Proof 14 *Given that*

$$\|Cf(x, y)\| \leq \theta(\|x\|^p + \|y\|^p),$$

that is,

$$\|f(x + iy) + f(x - iy) - 2f(x) + 2f(y)\| \leq \theta(\|x\|^p + \|y\|^p).$$

Putting $y = -ix$, in the last inequality, we find

$$\|f(2x) - 4f(x)\| \leq 2\theta\|x\|^p. \tag{1.80}$$

Replacing x by $x/2$ in (1.80), we get

$$\|f(x) - 4f(x/2)\| \leq \frac{2\theta}{2^p}\|x\|^p, \quad \textit{for}\ x \in G.$$

Hence,

$$\begin{aligned}\|4^n f(x/2^n) - 4^m f(x/2^m)\| &= \|4^n f(x/2^n) - 4^m f(x/2^m) + 4^{m+1} f(x/2^{m+1}) \\ &- 4^{m+1} f(x/2^{m+1}) + \cdots + 4^{n-1} f(x/2^{n-1}) \\ &- 4^{n-1} f(x/2^{n-1})\| \\ &\leq \|4^m f(x/2^m) - 4^{m+1} f(x/2^{m+1})\| + \cdots \\ &+ \|4^{n-1} f(x/2^{n-1}) - 4^n f(x/2^n)\| \\ &\leq \sum_{j=m}^{n-1} \|4^j f(x/2^j) - 4^{j+1} f(x/2^{j+1})\|\end{aligned}$$

$$
\begin{aligned}
&= \sum_{j=m}^{n-1} 4^j \|f(x/2^j) - 4f(x/2^{j+1})\| \\
&= \sum_{j=m}^{n-1} \frac{4^j 2\theta}{2^p 2^{pj}} \|x\|^p = \sum_{j=m}^{n-1} \frac{2.4^j \theta}{2^{pj+p}} \|x\|^p \\
&= 2\theta \|x\|^p (4^m/2^{p(m+1)} + \cdots + 4^{n-1}/2^{pn}) \\
&= 2\theta 2^p \|x\|^p 4^m/2^{p(m+1)} \frac{1-(1/2^{p-2})^{n-m}}{2^p - 4} \qquad (1.81)
\end{aligned}
$$

for all non-negative integers n *and* m *with* $n > m$. *Taking limit* $n \to \infty$ *and keeping* m *as fixed in (1.81), it follows that* $\{4^n f(x/2^n)\}$ *is a Cauchy sequence for all* $x \in X$. *Let there be a mapping* $Q : X \to Y$ *such that*

$$Q(x) = \lim_{n\to\infty} 4^n f(x/2^n) \ \ for \ \ all \ \ x \in X.$$

From (1.75), it follows that

$$\|CQ(x,y)\| = \lim_{n\to\infty} 4^n \|Cf(x/2^n, y/2^n)\| \le \lim_{n\to\infty} \frac{4^n \theta}{2^{pn}} (\|x\|^p + \|y\|^p) = 0,$$

for all $x, y \in X$. *So* $CQ(x,y) = 0$. *Hence,* $Q : X \to Y$ *is quadratic. Taking* $m = 0$ *and* $n \to \infty$ *in (1.81), we obtain*

$$
\begin{aligned}
\lim_{n\to\infty} \|f(x) - 4^n f(x/2^n)\| &\le \lim\nolimits_{n\to\infty} \textstyle\sum_{j=0}^{n-1} \frac{2.4^j \theta}{2^{pj+p}} \|x\|^p \\
&= \tfrac{2\theta}{2^p} \tfrac{1}{1-4/2^p} \|x\|^p = \tfrac{2\theta}{2^p} \tfrac{2^p}{2^p-4} \|x\|^p,
\end{aligned}
$$

that is,

$$\|f(x) - Q(x)\| \le \frac{2\theta}{2^p - 4} \|x\|^p.$$

The uniqueness of Q *is similar to Theorem 8. This completes the proof of the theorem.*

Theorem 10 *Let* $p < 1$ *and* θ *be a positive real number, and let* $f : X \to Y$ *be a mapping satisfying* $f(ix) = -f(x)$ *and*

$$\|Cf(x,y)\| \le \theta.\|x\|^p.\|y\|^p \qquad (1.82)$$

for all $x, y \in X$. *Then there exists a unique quadratic mapping* $Q : X \to Y$ *such that*

$$\|f(x) - Q(x)\| \le \frac{\theta}{4 - 4^p} \|x\|^{2p} \ \ for \ \ all \ \ x \in X. \qquad (1.83)$$

Proof 15 *Putting* $y = -ix$ *in* (1.82), *we find that*

$$\|f(2x) - 4f(2x)\| \le \theta \|x\|^p.\| - ix\|^p = \theta \|x\|^{2p}, \qquad (1.84)$$

for all $x \in X$*. Hence,*

$$\|f(x) - 1/4f(2x)\| \leq (\theta/4)\|x\|^{2p},$$

for all $x \in X$*. Now,*

$$\begin{aligned}
&\|1/4^n f(2^n x) - 1/4^m f(2^m x)\| \\
&= \|1/4^n f(2^n x) - 1/4^m f(2^m x) + 1/4^{m+1} f(2^{m+1} x) \\
&+ \cdots - 1/4^{n-1} f(2^{n-1} x) + 1/4^{n-1} f(2^{n-1} x)\| \\
&\leq \|\frac{1}{4^m} f(2^m x) - \frac{1}{4^{m+1}} f(2^{m+1} x)\| + \cdots \\
&+ \|\frac{1}{4^{n-1}} f(2^{n-1} x) - \frac{1}{4^n} f(2^n x)\| \\
&= \sum_{k=m}^{n-1} \|1/4^k f(2^k x) - 1/4^{k+1} f(2^{k+1} x)\| \\
&= \sum_{k=m}^{n-1} \frac{1}{4^k} \|f(2^k x) - 1/4 f(2^{k+1} x)\| \\
&\leq \sum_{k=m}^{n-1} \frac{1}{4^k} \frac{\theta}{4} \|2^k x\|^{2p} = \sum_{k=m}^{n-1} \frac{4^{pk}\theta}{4^{k+1}} \|x\|^{2p} \qquad (1.85)
\end{aligned}$$

for all non-negative integers n *and* m *with* $n > m$*. Taking limit* $n \to \infty$ *and* m *as fixed in* (1.85)*, the right side converges and hence* $\{4^n f(x/2^n)\}$ *is a Cauchy sequence for all* $x \in X$*. Let there be a mapping* $Q : X \to Y$ *such that* $Q(x) = \lim_{n\to\infty} f(2^n x)$*, for all* $x \in X$*. From* (1.82)*,*

$$\begin{aligned}
\|CQ(x,y)\| &= \lim_{n\to\infty} \frac{1}{4^n} \|Cf(2^n x, 2^n y)\| \\
&\leq \lim_{n\to\infty} \frac{4^{pn}\theta}{4^n} \|x\|^p \|y\|^p = 0
\end{aligned}$$

for all $x, y \in X$*. So* $CQ(x,y) = 0$*. By Proposition 1, the mapping* $Q : X \to Y$ *is quadratic. Again taking* $m = 0$ *and limit* $n \to \infty$ *in* (1.85)*, we find* (1.83)*. Uniqueness of* Q *is similar to that of Theorem 8.*

Theorem 11 *Let* $p > 1$ *and* θ *be positive real numbers, and let* $F : X \to Y$ *be a mapping satisfying*

$$\|Cf(x,y)\| \leq \theta.\|x\|^p.\|y\|^p$$

and $f(ix) = -f(x)$ *for all* $x \in X$*. Then there exists a unique quadratic mapping* $Q : X \to Y$ *such that*

$$\|f(x) - Q(x)\| \leq \frac{\theta}{4^p - 4} \|x\|^{2p} \qquad (1.86)$$

for all $x \in X$*.*

Proof 16 *From (1.84), it follows that*

$$\|f(x) - 4f(x/2)\| \leq \frac{\theta}{4^p}\|x\|^{2p}$$

for all $x \in X$. Therefore,

$$\|4^n f(x/2^n) - 4^m f(x/2^m)\| \leq \sum_{j=1}^{m-1} \frac{4^j \theta}{4^{pj+p}}\|x\|^{2p} \tag{1.87}$$

for all non-negative n and m with $n > m$ and all $x \in X$. As $n \to \infty$ in (1.87), it follows that $\{4^n f(x/2^n)\}$ is a Cauchy sequence for all $x \in X$. If $Q : X \to Y$ is defined by

$$\lim_{n\to\infty} 4^n f(x/2^n) = Q(x), \;\; \forall \; x \in X,$$

then

$$\|CQ(x,y)\| = \lim_{n\to\infty} 4^n \|Cf(x/2^n, y/2^n)\| \leq \lim_{n\to\infty} \frac{4^n\theta}{4^{pn}}.\|x\|^p.\|y\|^p = 0$$

for all $x \in X$. Thus $CQ(x,y) = 0$. By Proposition 1, the mapping $Q : X \to Y$ is quadratic. Also, putting $m = 0$ and taking limit $n \to \infty$ in (1.87),we find (1.86). Uniqueness of Q is similar to that of Theorem 8. This completes the proof of the theorem.

Let X be a normed vector space and Y be a Banach space. There is a mapping $f : X \to Y$ such that

$$Df(x,y) = f(x+iy) - f(ix+y) - 2f(x) + 2f(y)$$

for all $x, y \in X$. In the following we discuss the generalized Hyers-Ulam stability of the quadratic functional equation $Df(x,y) = 0$ by applying the fixed point method.

Theorem 12 *Let $f : X \to Y$ be a mapping with $f(ix) = -f(x)$ for all $x \in X$ for which there exists a function $\varphi : X \times X \to [0, \infty)$ such that*

$$\sum_{j=0}^{\infty} 4^{-j}\varphi(2^j x, 2^j y) < \infty \tag{1.88}$$

and

$$\|Df(x,y)\| \leq \varphi(x,y) \tag{1.89}$$

for all $x, y \in X$. If there exists an $L < 1$ such that $\varphi(x, -ix) \leq 4L\varphi(x/2, -ix/2)$ for all $x \in X$, then there exists a unique quadratic mapping $Q : X \to Y$ satisfying

$$f(x+iy) - f(ix+y) = 2f(x) - 2f(y)$$

and

$$\|f(x) - Q(x)\| \leq \frac{1}{4-4L}\varphi(x, -ix) \tag{1.90}$$

for all $x \in X$.

Proof 17 *Because of* $f(ix) = -f(x)$ *for all* $x \in X$, $f(0) = 0$.

$$f(-x) = f(i^2x) = -f(ix) = f(x)$$

for all $x \in X$. *Therefore,*

$$\begin{aligned}
Df(x,y) &= f(x+iy) - f(ix+y) - 2f(x) + 2f(y)\\
&= f(x+iy) + f(i(ix+y)) - 2f(x) + 2f(y)\\
&= f(x+iy) + f(-(x-iy)) - 2f(x) + 2f(y)\\
&= f(x+iy) + f(x-iy) - 2f(x) + 2f(y)\\
&= Cf(x,y), \ \ \forall \ x,y \in X.
\end{aligned}$$

Let us define the set $S := \{g : X \to Y\}$ *and develop the generalized metric on* S *as*

$$d(g,h) = inf\{K \in R_+ : \|g(x) - h(x)\| \le K\varphi(x,-ix), \ \ \forall x \in X\}.$$

Hence d is finite and Y *is a Banach space. Let* $\{g_n\}$ *be a Cauchy sequence. Then* $\|g_n - g\| \to 0$ *as* $n \to \infty$. *Now* $d(g_n, g) < \infty$. *So,* $g \in S$. *Therefore,* (S,d) *is complete.*
Define the linear mapping $J : S \to S$ *such that*

$$Jg(x) = \frac{1}{4}g(2x), \ \ \forall \ x \in S.$$

By Theorem 2 and for $g,h \in S$ *we have* $d(Jg,Jh) \le Ld(g,h)$ *for all* $g,h \in S$, *that is,* J *is a strictly contractive self-mapping of* S *with the Lipschitz constant* L, *and it follows that* $d(f,Jf) \le L$ *for all* $g,h \in S$. *Putting* $y = -ix$ *in (1.89), we find*

$$\|f(x+x) - f(0) - 2f(x) + 2f(-ix)\| \le \varphi(x,-ix),$$

that is,

$$\|f(2x) - 4f(x)\| \le \varphi(x,-ix),$$

for all $x \in X$. *Therefore,*

$$\|f(x) - 1/4f(2x)\| \le 1/4\varphi(x,-ix)$$

for all $x \in X$. *Thus* $d(f,Jf) \le 1/4$. *By Theorem 1, there exists a mapping* $Q : X \to Y$ *such that*

(1) Q *is a fixed point of* J, *that is,*

$$Q(2x) = 4Q(x), \ \ for \ \ all \ \ x \in X. \tag{1.91}$$

The mapping Q *is a unique fixed point of* J *in the set*

$$M = \{g \in S : d(f,g) < \infty\}.$$

It follows that Q is a unique mapping satisfying (1.91) such that there exists $K \in (0, \infty)$ satisfying

$$\|f(x) - Q(x)\| \leq K\varphi(x, -ix)$$

for all $x \in X$.

(2) $d(J^n f, Q) \to 0$ as $n \to \infty$. Therefore

$$\lim_{n\to\infty} \frac{f(2^n x)}{4^n} = Q(x) \;\; for \;\; all \;\; x \in X. \tag{1.92}$$

(3) $d(f, Q) \leq \frac{1}{1-L} d(f, Jf)$ which implies that

$$d(f, Q) \leq \frac{1}{4 - 4L}$$

which is (1.90). Now,

$$\|DQ(x, y)\| = \lim_{n\to\infty} (1/4^n)\|Df(2^n x, 2^n y) \leq \lim_{n\to\infty} (1/4^n)\varphi(2^n x, 2^n y) = 0$$

for all $x, y \in X$ due to (1.88), (1.89) and (1.92).

Hence $DQ(x, y) = 0$ for all $x, y \in X$. Clearly, $Q(ix) = -Q(x)$ is equivalent to $f(ix) = -f(x)$ for all $x \in X$. By Proposition 2, the mapping $Q : X \to Y$ is quadratic. This completes the proof of the theorem.

Corollary 10 *Let $p < 2$ and θ be positive real numbers, and let $f : X \to Y$ be a mapping satisfying $f(ix) = -f(x)$ and*

$$\|Df(x, y)\| \leq \theta(\|x\|^p + \|y\|^p) \tag{1.93}$$

for all $x, y \in X$. Then there exists a unique quadratic mapping $Q : X \to Y$ satisfying $f(x + iy) - f(ix + y) = 2f(x) - 2f(y)$ and

$$\|f(x) - Q(x)\| \leq \frac{2\theta}{4 - 2^p}\|x\|^p$$

for all $x \in X$.

Proof 18 *Let $\varphi(x, y) = \theta(\|x\|^p + \|y\|^p)$ for all $x, y \in X$. Replacing y by $-ix$, we obtain*

$$\begin{aligned} \varphi(x, -ix) &= \theta(\|x\|^p + \|-ix\|^p) \\ &= \theta(\|x\|^p + \|x\|^p) = 2\theta\|x\|^p. \end{aligned}$$

Therefore, as in Theorem 12

$$\|f(x) - (1/4)f(2x)\| \leq (1/2)\theta\|x\|^p.$$

Since $f(x+iy)-f(ix+y)-2f(x)+2f(y)=\theta(\|x\|^p+\|y\|^p)$, then replacing $2x$ by x and $y=0$, we find

$$f(2x)-f(i2x)-2f(2x)=\theta\|2x\|^p$$

implies that

$$\|4f(x)-f(2x)\|\le 2^p\theta\|x\|^p,$$

where we have used the fact that f is quadratic due to Proposition 2. Therefore,

$$\|f(x)-(1/4)f(2x)\|\le 2^{p-2}\theta\|x\|^p.$$

For $L=2^{p-2}$ and by Theorem 12, it follows that

$$\|f(x)-Q(x)\|\le \frac{2\theta}{4-2^p}\|x\|^p$$

for all $x\in X$. Q is quadratic as proved in Theorem 12. This completes the proof of the corollary.

Theorem 13 *Let $f:X\to Y$ be a mapping with $f(ix)=-f(x)$ for all $x\in X$ for which there exists a function $\varphi:X\times X\to[0,\infty)$ satisfying $\|Df(x,y)\|\le\varphi(x,y)$ such that*

$$\sum_{j=0}^{\infty}4^j\varphi(\frac{x}{2^j},\frac{y}{2^j})<\infty \tag{1.94}$$

for all $x,y\in X$. If there exists an $L<1$ such that $\varphi(x,-ix)\le\frac{1}{4}L\varphi(2x,-2ix)$ for all $x\in X$, then there exists a unique quadratic mapping $Q:X\to Y$ satisfying $f(x+iy)-f(ix+y)=2f(x)-2f(y)$ and

$$\|f(x)-Q(x)\|\le\frac{L}{4-4L}\varphi(x,-ix) \tag{1.95}$$

for all $x\in X$.

Proof 19 *Define the linear mapping $J:S\to S$ such that*

$$Jg(x)=4g(x/2),\quad for\ \ all\ \ x\in S.$$

From Theorem 12, we have that $\|f(2x)-4f(x)\|\le\varphi(x,-ix)$. Hence,

$$\|f(x)-4f(x/2)\|\le\varphi(x/2,-ix/2)\le\frac{L}{4}\varphi(x,-ix)$$

for all $x\in X$. Thus $d(f,Jf)\le\frac{L}{4}$ by Theorem 2. So by Theorem 1, there exists a mapping $Q:X\to Y$ such that

(1) Q is a fixed point of J and

$$Q(2x) = 4Q(x) \quad for \;\; all \;\; x \in X.$$

The mapping Q is a unique fixed point of J in the set

$$M = \{g \in S : d(f, g) < \infty\}.$$

It follows that Q is a unique mapping satisfying $Q(2x) = 4Q(x)$ such that there exists $K \in (0, \infty)$ satisfying

$$\|f(x) - Q(x)\| \leq K\varphi(x, -ix)$$

for all $x \in X$.

(2) $d(J^n f, Q) \to 0$ as $n \to \infty$. Therefore,

$$\lim_{n\to\infty} 4^n f(x/2^n) = Q(x), \quad for \;\; all \;\; x \in X.$$

(3) $d(f, Q) \leq \frac{1}{1-L} d(f, Jf)$ implies that

$$d(f, Q) \leq \frac{L}{4 - 4L}.$$

Therefore, (1.95) is true.This completes the proof of the theorem.

1.2 Stability of Some Functional Equations on Abelian Groups

This section deals with the Hyers-Ulam stability of the quadratic functional

$$f(x+y+z)+f(x-y)+f(y-z)+f(x-z) = 3f(x)+3f(y)+3f(z) \quad (1.96)$$

on abelian groups and the Hyers-Ulam stability of the

$$f(rx + sy) = \frac{r+s}{2} f(x+y) + \frac{r-s}{2} f(x-y), \qquad (1.97)$$

where $r, s \in \mathbb{R}$ and $r \neq \pm s$ over a unital C^*-algebra. It is well known that a function $f : E_1 \to E_2$ between vector spaces is quadratic if and only if there exists a unique symmetric function $B : E_1 \times E_1 \to E_2$, which is additive in x for each fixed y, such that $f(x) = B(x, x)$ for any $x \in E_1$.

A commutative semigroup $(X, \|.\|)$ with zero and metric d satisfying the following conditions:

1. Cancellation law: $a + c = b + c$ implies $a = b$ for all $a, b, c \in X$.

2. A multiplication by nonnegative real scalars is defined on X as follows

$$\alpha(a + b) = \alpha a + \alpha b, \quad \alpha a + \beta a = (\alpha + \beta)a,$$

$$\alpha(\beta a) = (\alpha\beta)a, \quad 1a = a,$$

for all $a, b \in X$ and $\alpha, \beta \in R_+$.

3. Metric d on X defined as
$d(x + y, x + z) = d(y, z)$ for all $x, y, z \in X$.
$d(tx, ty) = td(x, y)$ for all $x, y \in X$, $t \in R_+$
is called a quasi-normed space with the norm defined by $\|x\| := d(x, 0), x \in X$.

1.2.1 Stability of $f(x + y + z) + f(x - y) + f(y - z) + f(x - z) = 3f(x) + 3f(y) + 3f(z)$

Let E_1 be a group. Let $h : E_1 \times E_1 \times E_1 \to \mathbb{R}_+$ be a given function. We denote

$$H(x, y, z) := h(x, y, z) + h(x, 0, 0) + h(y, 0, 0) + h(z, 0, 0) + 2h(0, 0, 0)$$

$$\begin{aligned} K(x, y, z) := 3h(x, y, z) + h(x + y + z, 0, 0) + h(x - y, 0, 0) + h(y - z, 0, 0) \\ + h(x - z, 0, 0) + h(0, 0, 0), \end{aligned}$$

for all $x, y, z \in E_1$.

Lemma 1 *Let E_1 be a group and E_2 a quasi-normed space. If the functions $F, G : E_1 \to E_2$ fulfil the inequality*

$$\begin{aligned} &d[F(x + y + z) + F(x - y) + F(y - z) + F(x - z), \\ &G(x) + G(y) + G(z)] \leq h(x, y, z), \end{aligned} \tag{1.98}$$

for all $x, y, z \in E_1$, then we have

$$\begin{aligned} &d[F(x + y + z) + F(x - y) + F(y - z) + F(x - z) + 5F(0), \\ &3F(x) + 3F(y) + 3F(z)] \leq H(x, y, z) \end{aligned} \tag{1.99}$$

and

$$\begin{aligned} &d[G(x + y + z) + G(x - y) + G(y - z) + G(x - z) + 5G(0), \\ &3G(x) + 3G(y) + 3G(z)] \leq K(x, y, z) \end{aligned} \tag{1.100}$$

for all $x, y, z \in E_1$.

Proof 20 *Clearly,*

$$\begin{aligned}
d[&G(x+y+z)+G(x-y)+G(y-z)+G(x-z)+5G(0),\\
&3G(x)+3G(y)+3G(z)]\\
&\leq d[G(x+y+z)+G(x-y)+G(y-z)+G(x-z)+2G(0),\\
&3F(x+y+z)+G(x-y)+G(y-z)+G(x-z)+F(0)]\\
&+d[G(x-y)+2G(0),3F(x-y)+F(0)]\\
&+d[G(y-z)+2G(0),3F(y-z)+F(0)]\\
&+d[G(x-z)+2G(0),3F(x-z)+F(0)]\\
&+d[3F(x+y+z)+3F(x-y)+3F(y-z)+3F(x-z),\\
&3G(x)+3G(y)+3G(z)]+d[4F(0),3G(0)] \qquad (1.101)
\end{aligned}$$

Using (1.98), we obtain

$$\begin{aligned}
h(x+y+z,0,0) &\geq d[F(x+y+z)+F(x+y+z)+F(x+y+z)+F(0),\\
&G(x+y+z)+G(0)+G(0)]\\
&=d[3F(x+y+z)+F(0),G(x+y+z)+2G(0)], \qquad (1.102)
\end{aligned}$$

$$\begin{aligned}
h(x-y,0,0) &\geq d[F(x-y)+F(x-y)+F(0)+F(x-y),\\
&G(x-y)+G(0)+G(0)]\\
&=d[3F(x-y)+F(0),G(x-y)+2G(0)], \qquad (1.103)
\end{aligned}$$

$$\begin{aligned}
h(y-z,0,0) &\geq d[F(y-z)+F(y-z)+F(0)+F(y-z),\\
&G(y-z)+G(0)+G(0)]\\
&=d[3F(y-z)+F(0),G(y-z)+2G(0)], \qquad (1.104)
\end{aligned}$$

$$\begin{aligned}
h(x-z,0,0) &\geq d[F(x-z)+F(x-z)+F(x-z)+F(0),\\
&G(x-z)+G(0)+G(0)]\\
&=d[3F(x-z)+F(0),G(x-z)+2G(0)], \qquad (1.105)
\end{aligned}$$

and

$$\begin{aligned}
h(0,0,0) &\geq d[F(0)+F(0)+F(0)+F(0),G(0)+G(0)+G(0)]\\
&=d[4F(0),3G(0)]. \qquad (1.106)
\end{aligned}$$

Putting (1.102)–(1.106) *in* (1.101), *we get*

$$\begin{aligned}
d[&G(x+y+z)+G(x-y)+G(y-z)+G(x-z)+5G(0),\\
&3G(x)+3G(y)+3G(z)]\\
&\leq h(x+y+z,0,0)+h(x-y,0,0)\\
&+h(y-z,0,0)+h(x-z,0,0)+3h(x,y,z)+h(0,0,0)\\
&=K(x,y,z)
\end{aligned}$$

which is (1.100). *Clearly,*

$$
\begin{aligned}
&d[F(x+y+z)+F(x-y)+F(y-z)+F(x-z)+5F(0),\\
&3F(x)+3F(y)+3F(z)]\\
&\le d[3F(x)+F(0),G(x)+G(0)]+d[8F(0),6G(0)]\\
&+d[F(x+y+z)+F(x-y)+F(y-z)+F(x-z),G(x)+G(y)+G(z)]\\
&+d[3F(y)+F(0),G(y)+2G(0)]+d[3F(z)+F(0),G(z)+2G(0)]. \quad (1.107)
\end{aligned}
$$

Since,

$$
\begin{aligned}
h(x,0,0) &\ge d[F(x)+F(x)+F(0)+F(x),G(x)+G(0)+G(0)]\\
&= d[3F(x)+F(0),G(x)+2G(0)],
\end{aligned}
$$

$$
\begin{aligned}
h(y,0,0) &\ge d[F(y)+F(y)+F(0)+F(y),G(y)+G(0)+G(0)]\\
&= d[3F(y)+F(0),G(y)+2G(0)]
\end{aligned}
$$

and

$$
\begin{aligned}
h(z,0,0) &\ge d[F(z)+F(z)+F(0)+F(z),G(z)+G(0)+G(0)]\\
&= d[3F(z)+F(0),G(z)+2G(0)],
\end{aligned}
$$

then (1.107) becomes

$$
\begin{aligned}
d[F(x+y+z)&+F(x-y)+F(y-z)+F(x-z)+5F(0),\\
&3F(x)+3F(y)+3F(z)]\\
&\le h(x,y,z)+h(x,0,0)+h(y,0,0)+h(z,0,0)+2h(0,0,0)\\
&\le H(x,y,z)
\end{aligned}
$$

which is (1.99). This completes the proof of the lemma.

Lemma 2 *Let E_1 be a group and E_2 a quasi-normed space. If the functions $F,G : E_1 \to E_2$ satisfy the inequality*

$$d[F(x+y+z)+F(x-y)+F(y-z)+F(x-z),G(x)+G(y)+G(z)] \le h(x,y,z),$$

then

$$
\begin{aligned}
&d[F(k^n x)+(k^{2n}-1)F(0),k^{2n}F(x)]\\
&\qquad \le k^{2(n-1)}\sum_{i=1}^{k-1}\sum_{j=0}^{n-1} b_i H((k-i)k^j x,k^j x,0)k^{-2j} \qquad (1.108)
\end{aligned}
$$

and

$$
\begin{aligned}
&d[G(k^n x)+(k^{2n}-1)G(0),k^{2n}G(x)]\\
&\qquad \le k^{2(n-1)}\sum_{i=1}^{k-1}\sum_{j=0}^{n-1} b_i K((k-i)k^j x,k^j x,0)k^{-2j} \qquad (1.109)
\end{aligned}
$$

for all $x \in E_1$ and $n,k \in \mathbb{N}$, where $k \ge 2$, $b_i = 2b_{i-1}+b_{i-2}$,($b_1 = 1$, $b_2 = 2$).

Proof 21 *We prove the lemma by the method of induction. For $n = 1$ in (1.108), we obtain*

$$d[F(kx) + (k^2 - 1)F(0), k^2F(x)] \leq \sum_{i=1}^{k-1} b_i H((k-i)x, x, 0) \tag{1.110}$$

for all $x \in E_1$ and all integer $k \geq 2$. Putting $z = 0$ and $y = x$ in (1.99), it follows that

$$d[F(x+x)+F(x-x)+F(x)+F(x)+5F(0), 3F(x)+3F(x)+3F(0)] \leq H(x, x, 0)$$

and hence

$$d[F(2x) + 3F(0), 4F(x)] \leq H(x, x, 0)$$

for all $x \in E_1$ and all integers $k \geq 2$. Assume that (1.110) is true for $k \geq 2$ and all $x \in E_1$. Since,

$$\begin{aligned} d[F((k+1)x) + ((k+1)^2 - 1)F(0), (k+1)^2F(x)] \\ &\leq d[F((k+1)x) + F((k-1)x) + (k^2 + 2k)F(0), \\ &(k^2 + 2k - 2)F(0) + 2F(kx) + 2F(x)] \\ &+ d[2F(kx) + 2F(x) + (k^2 + 2k - 2)F(0), \\ &F((k-1)x) + (k+1)^2F(x)] \end{aligned} \tag{1.111}$$

and

$$\begin{aligned} H(kx, x, 0) &\geq d[F(kx + x + 0)1F(kx - x) + F(x - 0) + F(kx - 0) + 5F(0), \\ &3F(kx) + 3F(x) + 3F(0)] \\ &= d[F((k+1)x) + F((k-1)x) + F(x) + F(kx) + 5F(0), \\ &3F(kx) + 3F(x) + 3F(0)], \end{aligned}$$

then (1.111) reduces to

$$\begin{aligned} d[F((k+1)x) + ((k+1)^2 - 1)F(0), (k+1)^2F(x)] \leq H(kx, x, 0) \\ &+ d[2F(kx) + (k^2 + 2k - 2)F(0), \\ &F((k-1)x) + (k^2 + 2k - 1)F(x)]. \end{aligned} \tag{1.112}$$

Because

$$\begin{aligned} d[2F(kx) + 2(k^2 - 1)F(0), 2k^2F(x)] + d[(k-1)^2F(x), F((k-1)x) \\ &+ ((k-1)^2 - 1)F(0)] \\ &= 2F(kx) + 2(k^2 - 1)F(0) - 2k^2F(x) + (k-1)^2F(x) - F((k-1)x) \\ &- ((k-1)^2 - 1)F(0) \\ &= 2F(kx) + 2(k^2 - 1)F(0) - 2k^2F(x) + (k^2 + 1 - 2k)F(x) - F((k-1)x) \\ &- (k^2 - 2k)F(0) \end{aligned}$$

$$= 2F(kx) - (k^2 - 1 + 2k)F(x) - F((k-1)x) + (2k^2 - 2 - k^2 + 2k)F(0)$$
$$= 2F(kx) - (k^2 - 1 + 2k)F(x) - F((k-1)x) + (k^2 + 2k - 2)F(0)$$
$$= d[2F(kx) + (k^2 + 2k - 2)F(0), F((k-1)x) + (k^2 + 2k - 1)F(x)],$$

then (1.112) yields

$$\begin{aligned}
d[2F((k+1)x) &+ ((k+1)^2 - 1)F(0), (k+1)^2F(x)] \leq H(kx, x, 0) \\
&+ d[2F(kx) + 2(k^2 - 1)F(0), 2k^2F(x)] + d[(k-1)^2F(x), \\
&F((k-1)x) + ((k-1)^2 - 1)F(0)] \\
&\leq H(kx, x, 0) + \sum_{i=1}^{k-1} 2b_i H((k-i)x, x, 0) \\
&+ \sum_{i=1}^{k-1} b_i H((k-i-1)x, x, 0) \\
&= H(kx, x, 0) + 2b_1 H((k-1)x, x, 0) + \sum_{i=2}^{k-1} 2b_i H((k-i)x, x, 0) \\
&+ \sum_{i=1}^{k-1} b_i H((k-i-1)x, x, 0) \\
&= H(kx, x, 0) + \sum_{i=3}^{k} 2b_{i-1} H((k+1-i)x, x, 0) \\
&+ 2b_1 H((k-1)x, x, 0) + \sum_{i=3}^{k} b_{i-2} H((k+1-i)x, x, 0) \\
&= H(kx, x, 0) + 2b_1 H((k-1)x, x, 0)
\end{aligned}$$

$$\begin{aligned}
&+ \sum_{i=3}^{k} (2b_{i-1} + b_{i-2}) H((k+1-i)x, x, 0) \\
&= H(kx, x, 0) + 2b_1 H((k-1)x, x, 0) + \sum_{i=3}^{k} 2b_i H((k+1-i)x, x, 0) \\
&= b_1 H(kx, x, 0) + b_2 H((k-1)x, x, 0) + \sum_{i=3}^{k} 2b_i H((k+1-i)x, x, 0) \\
&= \sum_{i=1}^{k} b_i H((k+1-i)x, x, 0).
\end{aligned}$$

Hence, (1.110) holds for all $k \geq 2$ and all $x \in E_1$. Now, we assume that

(1.108) holds, for all $n = m \geq 1$ *and all* $x \in E_1$*, that is,*

$$d[F(k^m x) + (k^{2m} - 1)F(0), k^{2m}F(x)] \leq k^{2(m-1)} \sum_{i=1}^{k-1} \sum_{j=0}^{m-1} b_i H((k-i)k^j x, k^j x, 0)k^{-2j}.$$

For $n = m + 1$*,*

$$\begin{aligned} d[F(k^{m+1}x + (k^{2(m+1)} - 1)F(0), k^{2(m+1)}F(x)] &\leq d[F(k^{m+1}x) + (k^2 - 1)F(0), k^2 F(k^m x)] \\ &+ k^2 d[F(k^m x) + (k^{2m} - 1)F(0), k^{2m}F(x)] \\ &= \sum_{i=1}^{k-1} b_i H((k-i)k^m x, k^m x, 0) \\ &+ k^{2m} \sum_{i=1}^{k-1} \sum_{j=0}^{m-1} b_i H((k-i)k^j x, k^j x, 0)k^{-2j} \\ &= k^{2m} \sum_{i=1}^{k-1} \sum_{j=0}^{m} b_i H((k-i)k^j x, k^j x, 0)k^{-2j}. \end{aligned}$$

Thus, by the method of induction the lemma holds true.

Theorem 14 *Let* E_1 *be an abelian group and* E_2 *a Banach space. Let the function* $F, G : E_1 \to E_2$ *satisfy the following inequality:*

$$\|F(x+y+z)+F(x-y)+F(y-z)+F(x-z)-G(x)-G(y)-G(z)\| \leq h(x,y,z)$$

for all $x, y, z \in E_1$*. Suppose that the series*

$$\sum_{j=0}^{\infty} h((k-i)k^j x, k^j x, 0)k^{-2j} \quad \text{and} \quad \sum_{j=0}^{\infty} h(k^j x, 0, 0)k^{-2j} \tag{1.113}$$

are convergent for all $x \in E_1$ *and the condition,*

$$\liminf_{n\to\infty} h(k^n x, k^n y, k^n z)k^{-2n} = 0 \tag{1.114}$$

for all $x, y, z \in E_1$*, then there exists exactly one quadratic function* $A : E_1 \to E_2$ *such that*

$$\|A(x) + F(0) - F(x)\| \leq k^{-2} \sum_{i=1}^{k-1} \sum_{j=0}^{\infty} b_i H((k-i)k^j x, k^j x, 0)k^{-2j} \tag{1.115}$$

and

$$\|3A(x) + G(0) - G(x)\| \leq k^2 \sum_{i=1}^{k-1} \sum_{j=0}^{\infty} b_i K((k-i)k^j x, k^j x, 0)k^{-2j} \tag{1.116}$$

for all $x \in E_1$.
If moreover, E_1 is a linear and F is (i.e., $F^{-1}(U)$ is a Borel set in E_1 for every open set U in E_1) or the function $t \to F(tx)$ is continuous for each fixed $x \in E_1$, then

$$A(tx) = t^2 A(x) \tag{1.117}$$

for all $x \in E_1, t \in \mathbb{R}$.

Proof 22 *Define $A_n(x) := k^{-2n}F(k^n x)$ for all $x \in E_1$ and $n \in \mathbb{N}$. We claim that $A_n(x)$ is a , for all $x \in E_1$. For $n > r$ and $x \in E_1$,*

$$\begin{aligned}
\|A_n(x) - A_r(x)\| &= k^{-2n}\|F(k^n x) - k^{2(n-r)}F(k^r x)\| \\
&\leq k^{-2r}\|F(0)\| + k^{-2n}\|F(k^{n-r}k^r x) \\
&+ (k^{2(n-r)} - 1)F(0) - k^{2(n-r)}F(k^r x)\| \\
&= k^{-2r}\|F(0)\| + k^{-2}\sum_{i=1}^{k-1}\sum_{j=0}^{n-r-1} b_i H((k-i)k^{j+r}x, k^{j+r}x, 0)k^{-2(j+r)} \\
&= k^{-2r}\|F(0)\| + k^{-2}\sum_{i=1}^{k-1}\sum_{j=r}^{n-1} b_i H((k-i)k^j x, k^j x, 0)k^{-2j} \\
&\to 0 \quad as \quad r \to \infty.
\end{aligned}$$

Since E_2 is complete, then let

$$\lim_{n\to\infty} A_n(x) = A(x)$$

for all $x \in E_1$. Similarly, we can show that $k^{-2n}G(k^n x)$ is a Cauchy sequence. Now, we prove that,

$$3A(x) = \lim_{n\to\infty} k^{-2n}G(k^n x) \tag{1.118}$$

for all $x \in E_1$. Indeed,

$$\begin{aligned}
\|3A(x) - k^{-2n}G(k^n x)\| &\leq \|3A(x) - 3k^{-2n}F(k^n x)\| \\
&+ \|3k^{-2n}F(k^n x) + k^{-2n}F(0) - k^{-2n}G(k^n x) + 2k^{-2n}G(0)\| \\
&+ \|k^{-2n}G(k^n x) + 2k^{-2n}G(0) - k^{-2n}G(k^n x)\| + k^{-2n}\|F(0)\| \\
&\leq 3\|A(x) - A_n(x)\| + k^{-2n}h(k^n x, 0, 0) + 2k^{-2n}\|G(0)\| \\
&+ k^{-2n}\|F(0)\|
\end{aligned}$$

due to (1.98) and $\|x\| := d(x.o), x \in X$ implies that

$$\lim_{n\to\infty} \|3A(x) - k^{-2n}G(k^n x)\| = 0$$

for all $x \in E_1$. Next, we show that the function A is quadratic. Now,

$$\begin{aligned}
\|A_n(x+y+z) &+ A_n(x-y) + A_n(y-z) + A_n(x-z) \\
&- k^{-2n}G(k^n x) + k^{-2n}G(k^n y) + k^{-2n}G(k^n z)\| \\
&\leq k^{-2n}h(k^n x, k^n y, k^n z),
\end{aligned}$$

and so taking limit as $n \to \infty$, *we get*

$$A(x+y+z) + A(x-y) + A(y-z) + A(x-z) = 3A(x) + 3A(y) + 3A(z)$$

for all $x, y, z \in E_1$ *because of (1.118). From (1.108), it follows that*

$$d[k^{-2n}F(k^n x) + (1 - k^{-2n})F(0), F(x)] \\ \leq k^{-2} \sum_{i=1}^{k-1} \sum_{j=0}^{n-1} b_i H((k-i)k^j x, k^j x, 0) k^{-2j},$$

that is,

$$d[A_n(x) + (1 - k^{-2n})F(0), F(x)] \\ \leq k^{-2} \sum_{i=1}^{k-1} \sum_{j=0}^{n-1} b_i H((k-i)k^j x, k^j x, 0) k^{-2j}.$$

Upon taking limit as $n \to \infty$ *in the above inequality, we find that*

$$\|A(x) + F(0) - F(x)\| \leq k^{-2} \sum_{i=1}^{k-1} \sum_{j=0}^{\infty} b_i H((k-i)k^j x, k^j x, 0) k^{-2j}$$

which is (3.20). On the other hand, (1.109) gives

$$d[k^{-2n}G(k^n x) + (1 - k^{-2n})G(0), G(x)] \\ \leq k^{-2} \sum_{i=1}^{k-1} \sum_{j=0}^{n-1} b_i K((k-i)k^j x, k^j x, 0) k^{-2j},$$

and taking limit as $n \to \infty$, *we find*

$$\|3A(x) + G(0) - G(x)\| \leq k^{-2} \sum_{i=1}^{k-1} \sum_{j=0}^{\infty} b_i K((k-i)k^j x, k^j x, 0) k^{-2j}$$

which is (1.116). For uniqueness, assume that there exist two quadratic functions $C_m : E_1 \to E_2$, $m = 1, 2$ *such that*

$$\|C_m(x) + F(0) - F(x)\| \leq k^{-2} a_m \sum_{i=1}^{k-1} \sum_{j=0}^{\infty} b_i H((k-i)k^j x, k^j x, 0) k^{-2j},$$

where $x \in E_1$, $n \in \mathbb{N}$ *and* $a_m \geq 0$. *So,*

$$k^{-2n}\|C_1(k^n x) - C_2(k^n x)\| \leq k^{-2n}(\|C_1(k^n x) + F(0) - F(k^n x)\| \\ + \|F(k^n x) - C_2(k^n x) + F(0)\|)$$

$$\leq (a_1+a_2)k^{-2}\sum_{i=1}^{k-1}\sum_{j=0}^{\infty} b_i H((k-i)k^{j+n}x, k^{j+n}x, 0)k^{-2(j+n)}$$

$$= (a_1+a_2)k^{-2}\sum_{i=1}^{k-1}\sum_{j=n}^{\infty} b_i H((k-i)k^{j}x, k^{j}x, 0)k^{-2j}$$

$$\to 0 \quad as \quad n \to \infty.$$

Hence, $C_1(x) = C_2(x)$ for all $x \in E_1$.
To complete the proof of the theorem, let L be any continuous liner functional defined on the space E_2. Let $\varphi : \mathbb{R} \to \mathbb{R}$ be given by

$$\varphi(t) := L[A(tx)], \quad for \quad all \quad x \in E_1, t \in \mathbb{R},$$

where x is fixed. Then φ is quadratic function and moreover, as the pointwise limit of the sequence

$$\varphi_n(t) = k^{-2n}L[F(k^n tx)], \quad t \in \mathbb{R}$$

is also measurable and hence has the form $\varphi(t) = t^2\varphi(1)$ for $t \in \mathbb{R}$. Therefore, for all $t \in \mathbb{R}$ and all $x \in E_1$

$$L[A(tx)] = \varphi(t) = t^2\varphi(1) = L[t^2 A(x)]$$

which proves (1.117). This completes the proof of the theorem.

Lemma 3 *Let E_1 be a group divisible by k and E_2 a quasi-normed space. If $F, G : E_1 \to E_2$ satisfy the inequality (1.98), then*

$$d[F(x) + (k^{2n}-1)F(0), k^{2n}F(k^{-n}x)] \leq k^{-2}\sum_{i=1}^{k-1}\sum_{j=1}^{n} b_i H((k-i)k^{-j}x, k^{-j}x, 0)k^{2j} \tag{1.119}$$

and

$$d[G(x) + (k^{2n}-1)G(0), k^{2n}G(k^{-n}x)] \leq k^{-2}\sum_{i=1}^{k-1}\sum_{j=1}^{n} b_i K((k-i)k^{-j}x, k^{-j}x, 0)k^{2j} \tag{1.120}$$

for all $x \in E_1$ and $n, k \in \mathbb{N}$, where $k \geq 2$, $b_i = 2b_{i-1} + b_{i-2}$ $(b_1 = 1, b_2 = 2)$.

Proof 23 *Putting $x = k^{-n}t$ into (1.108), we get*

$$d[F(k^n.k^{-n}t) + (k^{2n}-1)F(0), k^{2n}F(k^{-n}t)] \leq k^{2(n-1)}\sum_{i=1}^{k-1}\sum_{j=0}^{n-1} b_i H((k-i)k^{j-n}t, k^{j-n}t, 0)k^{-2j}.$$

Let $n - j = m$*. When* $j = 0$*,* $m = n$ *and when* $j = n-1$*,* $m = 1$*. Hence,*

$$
\begin{aligned}
d[F(t) &+ (k^{2n}-1)F(0), k^{2n}F(k^{-n}t)] \\
&\leq k^{2(n-1)} \sum_{i=1}^{k-1} \sum_{m=1}^{n} b_i H((k-i)k^{-m}t, k^{-m}t, 0)k^{2(m-n)} \\
&= k^{2(n-1)}.k^{-2n} \sum_{i=1}^{k-1} \sum_{m=1}^{n} b_i H((k-i)k^{-m}t, k^{-m}t, 0)k^{2m} \\
&= k^{-2} \sum_{i=1}^{k-1} \sum_{m=1}^{n} b_i H((k-i)k^{-m}t, k^{-m}t, 0)k^{2m}
\end{aligned}
$$

which verifies (1.119)*. Similarly,* (1.120)*.*

Theorem 15 *Let* E_1 *be an abelian group divisible by* k*,* $k \geq 2$ *and* E_2 *a complete quasi-normed space. Suppose that the series*

$$\sum_{j=1}^{\infty} h((k-i)k^{-j}x, k^{-j}x, 0)k^{2j} \quad \text{and} \quad \sum_{j=1}^{\infty} h(k^{-j}x, 0, 0)k^{2j}$$

are convergent for all $x \in E_1$ *and the condition*

$$\liminf_{n\to\infty} h(k^{-n}x, k^{-n}y, k^{-n}z) = 0$$

for all $x, y, z \in E_1$*, then there exists exactly one quadratic function* $A : E_1 \to E_2$ *such that*

$$d[B(x), F(x)] \leq k^{-2} \sum_{i=1}^{k-1} \sum_{j=1}^{\infty} b_i H((k-i)k^{-j}x, k^{-j}x, 0)k^{2j} \tag{1.121}$$

and

$$d[B(x), F(x)] \leq k^{-2} \sum_{i=1}^{k-1} \sum_{j=1}^{\infty} b_i K((k-i)k^{-j}x, k^{-j}x, 0)k^{2j} \tag{1.122}$$

for all $x \in E_1$*.*

Proof 24 *Given that* $\sum_{j=1}^{\infty} h(k^{-j}x, 0, 0)k^{2j}$ *is convergent. Then it follows that* $\sum_{j=1}^{\infty} h(0,0,0)k^{2j}$ *is convergent, that is,* $\sum_{j=0}^{\infty} h(0,0,0)k^{2j}$*. Since*

$$\liminf_{n\to\infty} h(k^{-n}x, k^{-n}y, k^{-n}z) = 0,$$

then $h(0,0,0) = 0$*. So* $H(0,0,0) = 0$ *and* $4F(0) = 3G(0)$ *due to* (1.98)*. Thus,* $9d[F(0), F(0)] = 0$ *implies that* $F(0) = 0$ *and* $F(0) = G(0) = 0$*. Let*

$B_n(x) = k^{2n}F(k^{-n}x)$ *for all* $x \in E_1$ *and* $n \in \mathbb{N}$. *We claim that* $B_n(x)$ *is a Cauchy sequence for all* $x \in E_1$.

$$\begin{aligned}
\|B_n(x), B_m(x)\| &= \|k^{2n}F(k^{-n}x) - k^{2m}F(k^{-m}x)\| \\
&= k^{2n}\|F(k^{-n}x) - k^{2(m-n)}F(k^{-m}x)\| \\
&\leq k^{2m}\|F(0)\| + k^{2n}\|F(k^{m-n}.k^{-m}x) \\
&+ (k^{2(m-n)})F(0) - k^{2(m-n)}F(k^{-m}x)\| \\
&\leq k^{2m}\|F(0)\| \\
&+ k^{-2}\sum_{i=1}^{k-1}\sum_{j=0}^{m-n-1} b_i H((k-i)k^{-(j+m)}x, k^{-(j+m)}x, 0)k^{2(j+m)} \\
&= k^{2m}\|F(0)\| \\
&+ k^{-2}\sum_{i=1}^{k-1}\sum_{j=r}^{n-1} b_i H((k-i)k^{-j}x, k^{-j}x, 0)k^{2j} \\
&\to 0 \ \ as \ \ n \to \infty.
\end{aligned}$$

Since E_2 *is complete, then we let*

$$B(x) = \lim_{n\to\infty} B_n(x) \tag{1.123}$$

for all $x \in E_1$. *Now,*

$$\begin{aligned}
&\|3B(x) - k^{2n}G(k^{-n}x)\| \\
&\quad \leq \|3B(x) - 3.k^{2n}F(k^{-n}x)\| \\
&\quad + \|3.k^{2n}F(k^{-n}x) + k^{2n}F(0) - k^{2n}G(k^{-n}x) + 2k^{2n}G(0)\| \\
&\quad + \|k^{2n}G(k^{-n}x) + 2k^{2n}G(0) - k^{2n}G(k^{-n}x)\| + k^{2n}\|F(0)\| \\
&\quad \leq 3\|B(x) - B_n(x)\| + k^{2n}h(k^{-n}x, 0, 0) \\
&\quad + 2k^{2n}\|G(0)\| + k^{2n}\|F(0)\|
\end{aligned}$$

implies that $\lim_{n\to\infty}\|3B(x) - k^{2n}G(k^{-n}x)\| = 0$ *due to (1.123) and* $\liminf_{n\to\infty} h(k^{-n}x, k^{-n}y, k^{-n}z)k^{2n} = 0$. *Therefore,*

$$3B(x) = \lim_{n\to\infty} k^{2n}G(k^{-n}x) \tag{1.124}$$

for all $x \in E_1$. *Since* $d[B_n(x+y+z) + B_n(x-y) + B_n(y-z) + B_n(x-z), k^{2n}G(k^{-n}x)+k^{2n}G(k^{-n}y)+k^{2n}G(k^{-n}z)] \leq k^{2n}h(k^{-n}x, k^{-n}y, k^{-n}z)$, *then using (1.123) and (1.124), it follows that* $B(x+y+z) + B(x-y) + B(y-z) + B(x-z) = 3B(x) + 3B(y) + 3B(z)$ *for all* $x, y, z \in E_1$ *by taking limits of both sides as* $n \to \infty$. *Thus,* B *is a quadratic function. Using* $F(0) = 0$ *and* $B_n(x) = k^{2n}F(k^{-n})$ *in (1.119), we get (1.121) as* $n \to \infty$. *Similarly, (1.122). Uniqueness is similar to that of Theorem14. This completes the proof of the theorem.*

1.2.2 Stability of $f(rx+sy) = ((r+s)/2)f(x+y)+((r-s)/2)f(x-y)$

Throughout this section, we assume that A C^*-algebra is a unital with norm $\|.\|$, unit 1. Also we assume that X and Y are normed left A-module and Banach left A-module with norms $\|.\|_X$ and $\|.\|_Y$, respectively. Let $U(A)$ be the set of unitary elements in A, and let $r, s \in \mathbb{R}$ and $r \neq s$.
For a given mapping $f : X \to Y$, $u \in U(A)$ and a given $\mu \in \mathbb{C}$, we define $D_u f$, $D_\mu f : X^2 \to Y$ by

$$D_u f(x, y) := f(rux + suy) - \frac{r+s}{2}uf(x+y) - \frac{r-s}{2}uf(x-y) \tag{1.125}$$

$$D_\mu f(x, y) := f(r\mu x + s\mu y) - \frac{r+s}{2}\mu f(x+y) - \frac{r-s}{2}\mu f(x-y), \tag{1.126}$$

for all $x, y \in X$.

Proposition 3 *Let $L : X \to Y$ be a mapping with $L(0) = 0$ such that $D_u L(x, y) = 0$ for all $x, y \in X$, for all $u \in U(A)$. Then L is A-linear.*

Proof 25

$$D_u L(x, y) = L(rux + suy) - \frac{r+s}{2}uL(x+y) - \frac{r-s}{2}uL(x-y).$$

Given that $D_u L(x, y) = 0$. Hence, for additive mapping L

$$\begin{aligned}
L(rux + suy) &= \frac{r+s}{2}uL(x+y) + \frac{r-s}{2}uL(x-y) \\
&= (\frac{ru}{2} + \frac{su}{2})L(x+y) + (\frac{ru}{2} - \frac{su}{2})L(x-y) \\
&= \frac{ru}{2}(L(x+y) + L(x-y)) + \frac{su}{2}(L(x+y) - L(x-y)) \\
&= \frac{ru}{2}(L(x) + L(y) + L(x) - L(y)) \\
&+ \frac{su}{2}(L(x) + L(y) - L(x) + L(y)),
\end{aligned}$$

implies that

$$L(rux + suy) = ruL(x) + suL(y).$$

Thus, L is A-linear.

Corollary 11 *Let $L : X \to Y$ be such that $L(0) = 0$ satisfy $D_1 L(x, y) = 0$ for all $x, y \in X$. Then L is additive.*

Proof 26 *Since $D_1 L(rx + sy) = L(rx + sy) - \frac{r+s}{2}L(x+y) - \frac{r-s}{2}L(x-y)$, then*

$$L(rx + sy) = \frac{r+s}{2}L(x+y) + \frac{r-s}{2}L(x-y).$$

So, L is additive by (3).

Corollary 12 *A mapping $L : X \to Y$ such that $L(0) = 0$ satisfies $D_\mu L(x,y) = 0$ for all $x,y \in X$ and all $\mu \in T := \mu \in \mathbb{C} : |\mu| = 1$, if and only if L is $\mathbb{C}$-linear.*

Proof 27 *Given that*

$$D_\mu L(x,y) = L(r\mu x + s\mu y) - \frac{r+s}{2}\mu L(x+y) - \frac{r-s}{2}\mu L(x-y).$$

Since $D_\mu L(x,y) = 0$, then

$$\begin{aligned} L(r\mu x + s\mu y) &= \frac{r+s}{2}\mu L(x+y) + \frac{r-s}{2}\mu L(x-y) \\ &= \frac{r\mu}{2}(L(x+y) + L(x-y)) \\ &+ \frac{s\mu}{2}(L(x+y) - L(x-y)) \\ &= r\mu L(x) + s\mu L(y). \end{aligned}$$

So L is $\mathbb{C}$-linear.

Theorem 16 *Let $f; X \to Y$ be a mapping satisfying $f(0) = 0$ for which there exist a function $\varphi : X^2 \to [0,\infty)$ such that*

$$\lim_{k\to\infty} \frac{1}{2^k}\varphi(2^k x, 2^k y) = 0, \tag{1.127}$$

and

$$\begin{aligned} \tilde{\varphi}(x) = \sum_{k=0}^{\infty} \frac{1}{2^k} \Bigg\{ \varphi\left(\frac{2^{k+1}rx}{r^2-s^2}, \frac{-2^{k+1}sx}{r^2-s^2}\right) \\ + \varphi\left(\frac{2^k x}{r+s}, \frac{2^k x}{r+s}\right) + \varphi\left(\frac{2^k x}{r-s}, \frac{-2^k x}{r-s}\right)\Bigg\} < \infty, \end{aligned} \tag{1.128}$$

$$\|D_1 f(x,y)\|_Y \le \varphi(x,y) \tag{1.129}$$

for all $x,y \in X$. Then, there exists a unique additive mapping $L : X \to Y$ such that

$$\|f(x) - L(x)\|_Y \le \frac{1}{2}\tilde{\varphi}(x) \tag{1.130}$$

for all $x \in X$.

Proof 28 *Given that $\|D_1 f(x,y)\|_Y \le \varphi(x,y)$.So,*

$$\begin{aligned} &\left\| D_1 f(x,y) - D_1 f\left(\frac{x+y}{2}, \frac{x+y}{2}\right) - D_1 f\left(\frac{x-y}{2}, \frac{y-x}{2}\right)\right\|_Y \\ &\le \|D_1 f(x,y)\|_Y + \left\|D_1 f\left(\frac{x+y}{2}, \frac{x+y}{2}\right)\right\|_Y + \left\|D_1 f\left(\frac{x-y}{2}, \frac{y-x}{2}\right)\right\|_Y \\ &\le \varphi(x,y) + \varphi\left(\frac{x+y}{2}, \frac{x+y}{2}\right) + \varphi\left(\frac{x-y}{2}, \frac{y-x}{2}\right) \end{aligned}$$

for all $x, y \in X$. Taking $u = 1$ in (1.125), it follows that

$$\begin{aligned}
&\Bigg\| f(rx+sy) - \frac{r+s}{2} f(x+y) - \frac{r-s}{2} f(x-y) - f\left(\frac{r(x+y)}{2} + \frac{s(x+y)}{2}\right) \\
&+ \frac{r+s}{2} f\left(\frac{x+y}{2} + \frac{x+y}{2}\right) + \frac{r-s}{2} f\left(\frac{x+y}{2} - \frac{x+y}{2}\right) \\
&- f\left(r(\frac{x-y}{2}) + s(\frac{y-x}{2})\right) \\
&+ \frac{r+s}{2} f\left(\frac{x-y}{2} + \frac{y-x}{2}\right) + \frac{r-s}{2} f\left(\frac{x-y}{2} - (\frac{y-x}{2})\right) \Bigg\|_Y \\
&\le \left\| f(rx+sy) - \frac{r+s}{2} f(x+y) - \frac{r-s}{2} f(x-y) \right\|_Y \\
&+ \Bigg\| f\left(\frac{r(x+y)}{2} + \frac{s(x+y)}{2}\right) - \frac{r+s}{2} f\left(\frac{x+y}{2} + \frac{x+y}{2}\right) \\
&- \frac{r-s}{2} f\left(\frac{x+y}{2} - \frac{x+y}{2}\right) \Bigg\|_Y \\
&+ \Bigg\| f\left(r\left(\frac{x-y}{2}\right) + s\left(\frac{y-x}{2}\right)\right) - \frac{r+s}{2} f\left(\frac{x-y}{2} + \frac{y-x}{2}\right) \\
&- \frac{r-s}{2} f\left(\frac{x-y}{2} - \left(\frac{y-x}{2}\right)\right) \Bigg\|_Y \\
&\le \|D_1 f(x,y)\|_Y + \left\| D_1 f\left(\frac{x+y}{2}, \frac{x+y}{2}\right) \right\|_Y + \left\| D_1 f\left(\frac{x-y}{2}, \frac{y-x}{2}\right) \right\|_Y \\
&\le \varphi(x,y) + \varphi\left(\frac{x+y}{2}, \frac{x+y}{2}\right) + \varphi\left(\frac{x-y}{2}, \frac{y-x}{2}\right),
\end{aligned}$$

that is,

$$\begin{aligned}
\Bigg\| f(rx+sy) &- \frac{r+s}{2} f(x+y) - \frac{r-s}{2} f(x-y) - f\left(\frac{r+s}{2}(x+y)\right) \\
&+ \frac{r+s}{2} f(x+y) + \frac{r-s}{2} f(0) - f\left(\frac{r-s}{2}(x-y)\right) \\
&+ \frac{r+s}{2} f(0) + \frac{r-s}{2} f(x-y) \Bigg\|_Y \\
&\le \varphi(x,y) + \varphi\left(\frac{x+y}{2}, \frac{x+y}{2}\right) + \varphi\left(\frac{x-y}{2}, \frac{y-x}{2}\right).
\end{aligned}$$

Consequently,

$$\begin{aligned}
&\left\| f(rx+sy) - f\left(\frac{r+s}{2}(x+y)\right) - f\left(\frac{r-s}{2}(x-y)\right) \right\|_Y \\
&\le \varphi(x,y) + \varphi\left(\frac{x+y}{2}, \frac{x+y}{2}\right) + \varphi\left(\frac{x-y}{2}, \frac{y-x}{2}\right). \qquad (1.131)
\end{aligned}$$

Replacing x *by* $\frac{1}{r+s}x + \frac{1}{r-s}y$, y *by* $\frac{1}{r+s}x - \frac{1}{r-s}y$ *in (1.131), we obtain*

$$\begin{aligned}
&\Bigg\| f\left(r\left\{\frac{1}{r+s}x + \frac{1}{r-s}y\right\} + s\left\{\frac{1}{r+s}x - \frac{1}{r-s}y\right\}\right) \\
&- f\left(\frac{r+s}{2}\left\{\frac{1}{r+s}x + \frac{1}{r-s}y + \frac{1}{r+s}x - \frac{1}{r-s}y\right\}\right) \\
&- f\left(\frac{r-s}{2}\left\{\frac{1}{r+s}x + \frac{1}{r-s}y - \frac{1}{r+s}x + \frac{1}{r-s}y\right\}\right) \Bigg\|_Y \\
&\le \varphi\left(\frac{1}{r+s}x + \frac{1}{r-s}y, \frac{1}{r+s}x - \frac{1}{r-s}y\right) \\
&+ \varphi\left(\frac{1}{2}\left(\frac{x}{r+s} + \frac{y}{r-s} - \frac{x}{r+s} - \frac{y}{r-s}\right), \frac{1}{2}\left(\frac{x}{r+s} + \frac{y}{r-s} + \frac{x}{r+s} - \frac{y}{r-s}\right)\right) \\
&+ \varphi\left(\frac{1}{2}\left(\frac{x}{r+s} + \frac{y}{r-s} - \frac{x}{r+s} + \frac{y}{r-s}\right), \frac{1}{2}\left(\frac{x}{r+s} - \frac{y}{r-s} - \frac{x}{r+s} - \frac{y}{r-s}\right)\right),
\end{aligned}$$

that is,

$$\begin{aligned}
\|f(x+y) - f(x) - f(y)\|_Y& \\
&\le \varphi\left(\frac{x}{r+s} + \frac{y}{r-s}, \frac{x}{r+s} - \frac{y}{r-s}\right) \\
&+ \varphi\left(\frac{x}{r+s}, \frac{x}{r+s}\right) + \varphi\left(\frac{y}{r-s}, -\frac{y}{r-s}\right) \qquad (1.132)
\end{aligned}$$

for all $x, y \in X$. *Hence, for* $y = x$ *in (1.132), we get*

$$\begin{aligned}
\|f(2x) - 2f(x)\|_Y \le \varphi\left(\frac{2rx}{r^2 - s^2}, \frac{-2sx}{r^2 - s^2}\right) + \varphi\left(\frac{x}{r+s}, \frac{x}{r+s}\right)& \\
+ \varphi\left(\frac{x}{r-s}, \frac{-x}{r-s}\right)& \qquad (1.133)
\end{aligned}$$

for all $x \in X$. *Denote*

$$\Psi(x) = \varphi\left(\frac{2rx}{r^2 - s^2}, \frac{-2sx}{r^2 - s^2}\right) + \varphi\left(\frac{x}{r+s}, \frac{x}{r+s}\right) + \varphi\left(\frac{x}{r-s}, \frac{-x}{r-s}\right),$$

for all $x \in X$. *Then, it follows from (1.128) that*

$$\sum_{k=0}^{\infty} \left(\frac{1}{2^k}\right) \Psi(2^k x) = \tilde{\varphi}(x) < \infty \qquad (1.134)$$

for all $x \in X$. *Replacing* x *by* $2^k x$ *and dividing both sides of (1.133) by* 2^{k+1} *we get*

$$\|(\frac{1}{2^{k+1}})f(2^{k+1}x - (\frac{1}{2^k})f(2^k x))\|_Y \le (\frac{1}{2^{k+1}})\Psi(2^k x)$$

for all $x \in X$ and $k \in \mathbb{N}$. Since,

$$\begin{aligned}
&\|(\frac{1}{2^{k+1}})f(2^{k+1}x - (\frac{1}{2^m})f(2^m x))\|_Y \\
&\leq \sum_{l=m}^{k} \|(\frac{1}{2^{l+1}})f(2^{l+1}x) - \frac{1}{2^l}f(2^l x)\|_Y \\
&\leq \frac{1}{2}\sum_{l=m}^{k} \frac{1}{2^l}\Psi(2^l x) \qquad (1.135)
\end{aligned}$$

for all $x \in X$ and all integer $k \geq m \geq 0$, it follows from (1.134) that the sequence $\{\frac{f(2^k x)}{2^k}\}$ is a Cauchy sequence in Y for all $x \in X$. Define the mapping $L : X \to Y$ by $L(x) = \lim_{k\to\infty}\{\frac{f(2^k x)}{2^k}\}$ for all $x \in X$. Let $m = 0$ in (1.135) and taking limit as $k \to \infty$ in (1.135), we get

$$\begin{aligned}
\|D_1 L(x,y)\|_Y &= \|\lim_{k\to\infty} D_1 \frac{f(2^k x, 2^k y)}{2^k}\|_Y \\
&= \lim_{k\to\infty} \frac{1}{2^k}\|D_1 f(2^k x, 2^k y)\|_Y \\
&\leq \lim_{k\to\infty}\{\frac{1}{2^k}\}\varphi(2^k x, 2^k y) \\
&= 0 \quad for \ \ all \ \ x, y \in X,
\end{aligned}$$

where $D_1 L : X^2 \to Y$. Therefore, the mapping $L : X \to Y$ satisfies the equation

$$f(rx + sy) = \frac{r+s}{2}f(x+y) + \frac{r-s}{2}f(x-y)$$

and $L(0) = 0$. Hence by Proposition 3, L is a additive mapping. To prove the uniqueness of L . Let $L' : X \to Y$ be another additive mapping such that

$$\|f(x) - L(x)\|_Y \leq \frac{1}{2}\tilde{\varphi}(x).$$

Now,

$$\begin{aligned}
\|L(x) - L'(x)\|_Y &= \lim_{k\to\infty} \frac{1}{2^k}\|f(2^k x) - L'(2^k x)\|_Y \\
&\leq \frac{1}{2}\lim_{k\to\infty} \frac{1}{2^k}\sum_{l=0}^{\infty} \frac{1}{2^l}\Psi(2^{l+k}) \\
&= \frac{1}{2}\lim_{k\to\infty} \sum_{l=k}^{\infty} \frac{1}{2^l}\Psi(2^l x) \\
&= 0 \quad for \ \ all \ \ x \in X.
\end{aligned}$$

So $L(x) = L'(x)$ for all $x \in X$. This completes the proof of the theorem.

Theorem 17 *Let $f : X \to Y$ be a mapping satisfying $f(0) = 0$ for which there exists a function $\varphi : X^2 \to [0, 1)$ satisfying (1.127), (1.128) and*

$$\|D_u f(x, y)\| \le \varphi(x, y)$$

for all $x, y \in X$ and all $u \in U(A)$. Then there exists a unique A-linear mapping $L : X \to Y$ satisfying (1.130) for all $x \in X$.

Proof 29 *Putting $u = 1$ in $\|D_u f(x, y)\| \le \varphi(x, y)$, we get*

$$\|D_1 f(x, y)\| \le \varphi(x, y).$$

From Theorem 16, it follows that there exists a unique additive mapping $L : X \to Y$ such that $\|f(x) - L(x, y)\|_Y \le \frac{1}{2}\widetilde{\varphi}(x)$ for all $x \in X$. This completes proof of the theorem.

Corollary 13 *Let δ, ε, p and q be non-negative real numbers such that $0 < p, q < 1$. Assume that a mapping $f : X \to Y$ with $f(0) = 0$ satisfies the inequality*

$$\|D_1 f(x, y)\|_Y \le \delta + \varepsilon(\|x\|_X^p + \|y\|_X^q),$$
$$(\|D_u f(x, y)\|_Y \le \delta + \varepsilon(\|x\|_X^p + \|y\|_X^q))$$

for all $x, y \in X$ (and all $u \in U(A)$). Then there exists a unique additive (A-linear) mapping $L : X \to Y$ such that

$$\begin{aligned}\|f(x) - L(x)\|_Y \le 3\delta &+ \frac{2^p|r|^p + |r+s|^p + |r-s|^p}{(2-2^p)|r^2-s^2|^p}\varepsilon|x|_X^p \\ &+ \frac{2^q|s|^q + |r+s|^q + |r-s|^q}{(2-2^q)|r^2-s^2|^q}\varepsilon|x|_X^q\end{aligned}$$

for all $x \in X$.

Proof 30 *Let us define $\varphi(x, y) = \delta + \varepsilon(\|x\|_X^p + \|y\|_X^q)$. So,*

$$\varphi(2^k x, 2^k y) = \delta + \varepsilon(\|2^k x\|_X^p + \|2^k y\|_X^q).$$

Therefore,

$$\begin{aligned}\lim_{k\to\infty} \frac{\varphi(2^k x, 2^k y)}{2^k} &= \lim_{k\to\infty} \frac{1}{2^k}(\delta + \varepsilon(\|2^k x\|_X^p + \|2^k y\|_X^q)) \\ &= 0.\end{aligned}$$

Now,

$$\begin{aligned}&\sum_{k=0}^{\infty} \frac{1}{2^k}\left\{\varphi\left(\frac{2^{k+1}rx}{r^2-s^2}, \frac{-2^{k+1}sx}{r^2-s^2}\right) + \varphi\left(\frac{2^k x}{r+s}, \frac{2^k x}{r+s}\right) + \varphi\left(\frac{2^k x}{r-s}, \frac{-2^k x}{r-s}\right)\right\} \\ &= \sum_{k=0}^{\infty}\left\{\delta + \varepsilon\left(\left\|\frac{2^{k+1}rx}{r^2-s^2}\right\|_X^p + \left\|\frac{-2^{k+1}sx}{r^2-s^2}\right\|_X^q\right) + \delta\right. \\ &\quad \left. + \varepsilon\left(\left\|\frac{2^k x}{r+s}\right\|_X^p + \left\|\frac{2^k x}{r+s}\right\|_X^q\right) + \delta + \varepsilon\left(\left\|\frac{2^k x}{r-s}\right\|_X^p + \left\|\frac{-2^k x}{r-s}\right\|_X^q\right)\right\}.\end{aligned}$$

Then,

$$\sum_{k=0}^{\infty}\frac{1}{2^k}\left(\left|\frac{2^{k+1}r}{r^2-s^2}\right|^p+\frac{2^{kp}|r-s|^p+2^{kp}|r+s|^p}{|r^2-s^2|^p}\right)\varepsilon\|x\|_X^p$$
$$=\sum_{k=0}^{\infty}\frac{1}{2^k}\left(\frac{2^{(k+1)p}|r|^p+2^{kp}(|r-s|^p+|r+s|^p)}{|r^2-s^2|^p}\right)\varepsilon\|x\|_X^p$$
$$=\sum_{k=0}^{\infty}\frac{2^{kp}}{2^k}\left(\frac{2^p|r|^p+|r-s|^p+|r+s|^p}{|r^2-s^2|^p}\right)\varepsilon\|x\|_X^p.$$

Similarly,

$$\sum_{k=0}^{\infty}\frac{1}{2^k}\left(\left|\frac{2^{k+1}r}{r^2-s^2}\right|^q+\frac{2^{kq}|r-s|^q+2^{kq}|r+s|^q}{|r^2-s^2|^q}\right)\varepsilon\|x\|_X^q$$
$$=\sum_{k=0}^{\infty}\frac{2^{kq}}{2^k}\left(\frac{2^q|s|^q+|r-s|^q+|r+s|^q}{|r^2-s^2|^q}\right)\varepsilon\|x\|_X^q.$$

Hence from this two inequality and using Theorem 3.2.4, we obtain that

$$\|f(x)-L(x)\|_Y\le 3\delta+\frac{2^p|r|^p+|r-s|^p+|r+s|^p}{(2-2^p)|r^2-s^2|^p}\varepsilon\|x\|_X^p$$
$$+\frac{2^q|s|^q+|r-s|^q+|r+s|^q}{(2-2^q)|r^2-s^2|^p}\varepsilon\|x\|_X^q$$

for all $x\in X$.

Corollary 14 *Let δ,ε,p and q be non-negative real numbers such that $\lambda=p+q=1$ and $|r|\neq|r|^\lambda$. Assume that a mapping $f:X\to Y$ with $f(0)=0$ satisfies the inequality*

$$\|D_1f(x,y)\|_Y\le\varepsilon(\|x\|_X^p.\|y\|_X^q),$$
$$(\|D_uf(x,y)\|_Y\le\delta+\varepsilon(\|x\|_X^p.\|y\|_X^q))$$

for all $x,y\in X$ (and all $u\in U(A)$). Then f is additive (A-linear).

Proof 31 *Since*

$$\|D_1f(x,y)\|_Y\le\varepsilon(\|x\|_X^p.\|y\|_X^q),$$

then

$$\begin{aligned}\|D_1f(ax,ay)\|_Y&\le\varepsilon(\|ax\|_X^p.\|ay\|_X^q)\\&=a^{p+q}\varepsilon(\|x\|_X^p.\|y\|_X^q)\\&=a^{p+q}\|D_1f(x,y)\|_Y\\&=a\|D_1f(x,y)\|_Y\end{aligned}$$

due to $p+q=1$. This completes the proof of the corollary.

The study of stability problems for various functional equations originated from a talk given by S. M. Ulam in 1940. In that talk, Ulam discussed a number of important unsolved problems. Among such problems, a problem concerning the stability of functional equations is one of them. In 1941, Hyers [10] gave an answer to the problem.

Furthermore, the result of Hyers [10] has been generalized by Rassias [51]. After that many authors have extended the Ulam's stability problems to other functional equations and generalized Hyer's result in various directions. Thereafter, Ulam's stability problem for functional equations was replaced by stability of differential equations.

Definition 8 *The differential equation*

$$a_n(t)y^{(n)}(t) + a_{n-1}(t)y^{(n-1)}(t) + \cdots + a_1(t)y^{'}(t) + a_0(t)y(t) + h(t) = 0$$

has Hyers-Ulam stability, if for given $\epsilon > 0$*,* I *be an open interval and for any function* f *satisfying the differential inequality*

$$\left|a_n(t)y^{(n)}(t) + a_{n-1}(t)y^{(n-1)}(t) + \cdots + a_1(t)y^{'}(t) + a_0(t)y(t) + h(t)\right| \le \epsilon,$$

then there exists a solution f_0 *of the above differential equation such that* $|f(t) - f_0(t)| \le K(\epsilon)$ *and* $\lim_{\epsilon\to 0} K(\epsilon) = 0$*, for* $t \in I$*. If the preceding statement is also true when we replace* ϵ *and* $K(\epsilon)$ *by* $\phi(t)$ *and* $\psi(t)$ *respectively, where* $\phi, \psi\colon I \to [0,\infty)$ *are functions not depending on* f *and* f_0 *explicitly, then we say that the corresponding differential equation has the generalized Hyers-Ulam stability.*

S. M. Jung has investigated the Hyers-Ulam stability of linear differential equations of different classes. Keeping in view of the above definition of Hyers-Ulam stability for differential equations, we can view the corresponding definition of Hyers-Ulam stability for difference equations as follows:

Definition 9 *The difference equation*

$$\begin{aligned}&a_k(n)y(n+k) + a_{k-1}(n)y(n+k-1) + \cdots + a_1(n)y(n+1)\\&+a_0(n)y(n) + h(n) = 0\end{aligned}$$

has the Hyers-Ulam stability, if for given $\epsilon > 0$*,* I *be an open interval and for any real valued function* f *satisfying the inequality*

$$\begin{aligned}&|\ a_k(n)y(n+k) + a_{k-1}(n)y(n+k-1) + \cdots + a_1(n)y(n+1)\\&+a_0(n)y(n) + h(n)\ | \le \epsilon,\end{aligned}$$

then there exists a solution f_0 *of the above difference equation such that* $|\ f(n) - f_0(n)\ | \le K(\epsilon)$ *and* $\lim_{\epsilon\to 0} K(\epsilon) = 0$*, for* $n \in I \subset N(0) = \{0,1,2,3,\cdots\}$*.*

Because difference equations can be generated from differential equations by using the Euler's method, the above definitions exist simultaneously. Hence, the study of Hyers-Ulam stability for differential and difference equations is interesting.

1.3 Notes

In order to have an insight on Hyers-Ulam stability, Section 1.2 is devoted for some functional equations on Banach space while Section 1.3 deals with some more functional equations on abelian groups. For details, we refer [1], [2], [4], [7], [8], [12] and [14]. In addition, suggested references [17], [54] and [55] are provided for exclusive study of Hyers-Ulam stability of some other kind of functional equations.

Chapter 2

Stability of First Order Linear Differential Equations

The real exponential function $f(x) = e^x$ is the only nontrivial solution of the differential equation $f^{'} = f$ and the objective is to go through the the *Hyers-Ulam* stability of this equation, i.e. to solve for a given $\varepsilon > 0$ the inequality

$$|f^{'}(x) - f(x)| \leq \varepsilon \tag{2.1}$$

and also to study the related inequality:

$$\left| \frac{f(x+y)}{2} - \frac{f(y) - f(x)}{y - x} \right| \leq \varepsilon, \tag{2.2}$$

for all $x \neq y$. By using (2.1) and (2.2) several other inequalities can be solved. Here $I =$ any real interval and $\mathbb{R}^+$ for the set of all nonnegative real numbers. A function f will be termed if f satisfies the inequality

$$f\left(\frac{x+y}{2}\right) \geq \frac{f(x) + f(y)}{2}$$

and f will be said to be $k-$Lipschitz, whenever

$$|f(x) - f(y)| \leq k|x - y|,$$

$\forall\ x, y$ in the (convex) domain of f. A monotonic *Jensen* concave function $f : I \to \mathbb{R}$ has to be necessarily concave in the usual sense, that is, to satisfy the inequality

$$f\left(\lambda x + (1 - \lambda)y\right) \geq \lambda f(x) + (1 - \lambda)f(y),$$

for all $x, y \in I$ and all $\lambda \in [0, 1]$.

2.1 Stability of $f'(x) = f(x)$

Lemma 4 *Let $g : I \to \mathbb{R}$ be a differentiable function. Then:*

(i) *the inequality* $g(x) \leq g'(x)$ *holds for all* x *in* I *if and only if* g *can be represented in the form*

$$g(x) = i(x)e^x, x \in I, \tag{2.3}$$

where $i : I \to \mathbb{R}$ *is an arbitrary nondecreasing differentiable function;*

(ii) *the inequality* $g'(x) \leq g(x)$ *holds for all* x *in* I *if and only if* g *admits the representation*

$$g(x) = d(x)e^x, x \in I, \tag{2.4}$$

where $d : I \to \mathbb{R}$ *is an arbitrary nonincreasing differentiable function.*

Proof 32 *Let*

$$g(x) \leq g'(x), x \in I.$$

Then the function $i : I \to \mathbb{R}$ *is defined by*

$$i(x) = g(x)e^{-x}, x \in I,$$

is differentiable and can be written

$$i'(x) = g'(x)e^{-x} - g(x)e^{-x} = (g'(x) - g(x))e^{-x} \geq 0, \forall x \in I.$$

Hence i *is nondecreasing. Conversely, suppose that*

$$g(x) = i(x)e^x, x \in I.$$

Now,

$$g'(x) - g(x) = i'(x)e^x \geq 0.$$

Hence,

$$g(x) \leq g'(x).$$

So, part(i) is proved. For part (ii), let

$$g'(x) \leq g(x), x \in I.$$

Then the function $d : I \to \mathbb{R}$ *is defined by*

$$d(x) = g(x)e^{-x},$$

$x \in I$*, is differentiable and can be written*

$$d'(x) = g'(x)e^{-x} - g(x)e^{-x} = (g'(x) - g(x))e^{-x} \leq 0, \forall x \in I.$$

Hence d *is nonincreasing. Conversely, if*

$$g(x) = d(x)e^x, x \in I,$$

then

$$g'(x) - g(x) = d'(x)e^x \leq 0.$$

Hence,

$$g'(x) \leq g(x).$$

This completes the proof of the lemma.

Lemma 5 *For every real number $x \neq y$ the exponential function satisfies the inequalities*

$$e^{\frac{(x+y)}{2}} \leq \frac{e^y - e^x}{y - x} \leq \frac{e^x + e^y}{2}. \tag{2.5}$$

Proof 33 *For $t \geq 0$*

$$1 + \frac{t}{2} + \frac{t^2}{2^2 2!} + \frac{t^3}{2^3 3!} \cdots \leq 1 + \frac{t}{2!} + \frac{t^2}{3!} + \frac{t^3}{4!} + \cdots$$

implies that

$$1 + \left(\frac{t}{2}\right) + \frac{\left(\frac{t}{2}\right)^2}{2!} + \frac{\left(\frac{t}{2}\right)^3}{3!} \cdots \leq \frac{1}{t}\left(t + \frac{t^2}{2!} + \frac{t^3}{3!} + \frac{t^4}{4!} + \cdots\right).$$

Therefore,

$$1 + \left(\frac{t}{2}\right) + \frac{\left(\frac{t}{2}\right)^2}{2!} + \frac{\left(\frac{t}{2}\right)^3}{3!} \cdots \leq \frac{1}{t}\left(1 + t + \frac{t^2}{2!} + \frac{t^3}{3!} + \frac{t^4}{4!} + \cdots - 1\right)$$

and

$$e^{\frac{t}{2}} \leq \left(\frac{e^t - 1}{t}\right). \tag{2.6}$$

Again

$$1 + \frac{t}{2!} + \frac{t^2}{3!} + \frac{t^3}{4!} + \cdots \leq 1 + \frac{t}{2} + \frac{t^2}{2(2!)} + \frac{t^3}{2(3!)} + \cdots$$

implies that

$$\frac{1}{t}\left(t + \frac{t^2}{2!} + \frac{t^3}{3!} + \frac{t^4}{4!} + \cdots\right) \leq \frac{1}{2}\left(2 + t + \frac{t^2}{2!} + \frac{t^3}{3!} + \frac{t^4}{4!} + \cdots\right).$$

Hence,

$$\frac{1}{t}\left(1 + t + \frac{t^2}{2!} + \frac{t^3}{3!} + \frac{t^4}{4!} + \cdots - 1\right) \leq \frac{1}{2}\left(1 + t + \frac{t^2}{2!} + \frac{t^3}{3!} + \frac{t^4}{4!} + \cdots + 1\right),$$

that is,

$$\left(\frac{e^t - 1}{t}\right) \leq \left(\frac{e^t + 1}{2}\right). \tag{2.7}$$

From (2.6) and (2.7) we obtain

$$e^{\frac{t}{2}} \leq \frac{e^t - 1}{t} \leq \frac{e^t + 1}{2}. \tag{2.8}$$

Put $y - x = t$ in (2.8). Then

$$e^{\frac{y-x}{2}} \leq \frac{e^{y-x} - 1}{y - x} \leq \frac{e^{y-x} + 1}{2}.$$

Multiplying e^x *to the last expression, it follows that*

$$e^{\frac{x+y}{2}} \leq \frac{e^y - e^x}{y-x} \leq \frac{e^y + e^x}{2}.$$

This completes the proof of the lemma.

Lemma 6 *A function* $f : I \to \mathbb{R}^+$ *satisfies the inequality*

$$f\left(\frac{x+y}{2}\right) \leq \frac{f(y)-f(x)}{y-x} \tag{2.9}$$

for all $x \neq y$*, in* I*, if and only if* f *can be represented in the form* $f(x) = i(x)e^x$*, where* $i : I \to \mathbb{R}^+$ *is an arbitrary nondecreasing function.*

Proof 34 *Suppose that* f *satisfies* (2.9)*,* x *in* I*,* $h > 0$*, and* $x + h \in I$*. To prove that* f *can be represented in the form* $f(x) = i(x)e^x$*, where* $i : I \to \mathbb{R}^+$ *is an arbitrary nondecreasing function. Now,*

$$\frac{f(y)-f(x)}{y-x} \geq f\left(\frac{x+y}{2}\right)$$

implies that

$$f(y) - f(x) \geq (y-x)f\left(\frac{x+y}{2}\right), \forall y \geq x \tag{2.10}$$

or

$$f(y) - f(x) \leq (y-x)f\left(\frac{x+y}{2}\right), \forall y \leq x. \tag{2.11}$$

Hence from (2.10)*,* $f(y) - f(x) \geq 0, \forall y \geq x$*, that is,* $f(y) \geq f(x), \forall y \geq x$ *implies that* f *is a nondecreasing function and*

$$f(y) \geq f(x) + (y-x)f\left(\frac{x+y}{2}\right).$$

Put $x + h = y$ *in last inequality. Then*

$$f(x+h) \geq f(x) + ((x+h)-x)f\left(\frac{x+(x+h)}{2}\right),$$

that is,

$$f(x+h) \geq f(x) + hf\left(x + \frac{h}{2}\right).$$

Therefore,

$$f(x+h) \geq f(x) + hf\left(x + \frac{h}{2}\right) \geq f(x) + hf(x) = (1+h)f(x).$$

We claim that

$$f(x+ih) \geq (1+h)^i f(x),$$

for $x + ih \in I$ and $i \in \mathbb{N}$. Our claim is true for $i = 1$. Let's assume that it is true for $i = k$, that is, $f(x + kh) \geq (1 + h)^k f(x)$. For $i = k + 1$,

$$f(x + (k+1)h) = f((x + h) + kh) \geq (1 + h)^k f(x + h) \geq (1 + h)^k (1 + h) f(x),$$

implies that

$$f(x + (k+1)h) \geq (1 + h)^{k+1} f(x).$$

Hence, our claim is true. So, it is true for $\forall i \in \mathbb{N}$. Thus, for an arbitrarily fixed $n \in \mathbb{N}$, for every $x \leq y$ in I, we have

$$f(y) = f\left(x + n\frac{(y - x)}{n}\right) \geq \left(1 + \frac{y - x}{n}\right)^n f(x).$$

For $n \to \infty, f(y) \geq e^{y-x} f(x)$ implies that $f(y)e^{-y} \geq e^{-x} f(x), \forall y \geq x$. So, the function $i : I \to \mathbb{R}^+$ is defined by $i(x) = f(x)e^{-x}$ is nondecreasing.

Again from (2.11), $f(y) - f(x) \leq 0, \forall y \leq x$, that is, $f(y) \leq f(x), \forall y \leq x$, implies that f is a nondecreasing function and

$$f(y) \leq f(x) + (y - x) f\left(\frac{x + y}{2}\right).$$

Put $x - h = y$ in last inequality. Then

$$f(x - h) \leq f(x) + (x - h - x) f\left(\frac{x + x - h}{2}\right),$$

that is,

$$f(x - h) \leq f(x) - hf\left(x - \frac{h}{2}\right) \leq f(x) - hf(x - h)$$

implies that $f(x) \geq (1 + h)f(x - h)$, that is, $f(x - h) \leq \frac{f(x)}{1+h}$. We claim that

$$f(x - ih) \leq \frac{f(x)}{(1 + h)^i}$$

for $x - ih \in I$ and $i \in \mathbb{N}$. Our claim is true for $i = 1$. Let's assume that it is true for $i = k$, that is, $f(x - kh) \leq \frac{f(x)}{(1+h)^k}$. For $i = k + 1$,

$$f(x - (k+1)h) = f((x - h) - kh) \leq \frac{f(x - h)}{(1 + h)^k} \leq \frac{f(x)}{(1 + h)(1 + h)^k}$$

implies that

$$f(x - (k+1)h) \leq \frac{f(x)}{(1 + h)^{k+1}}.$$

Hence, our claim holds. So, it is true $\forall i \in \mathbb{N}$. Thus, for an arbitrarily fixed $n \in \mathbb{N}$ and for every $y \leq x$ in I, we have

$$f(y) = f\left(x + n\frac{(y - x)}{n}\right) = f\left(x - n\frac{(x - y)}{n}\right) \leq \frac{f(x)}{\left(1 + \frac{x-y}{n}\right)^n}.$$

For $n \to \infty$, $f(y) \leq \frac{f(x)}{e^{x-y}}$, *that is,* $e^{x-y} f(y) \leq f(x)$ *implies that* $e^{-y} f(y) \leq e^{-x} f(x), \forall y \leq x$. *So, the function* $i : I \to \mathbb{R}^+$, *is defined by* $i(x) = f(x) e^{-x}$, *is nondecreasing.*

Conversly, let $f(x) = i(x) e^x$, $x \in I$ *with* $i : I \to \mathbb{R}^+$ *is nondecreasing. To prove that*

$$f\left(\frac{x+y}{2}\right) \leq \frac{f(y) - f(x)}{y - x}.$$

For $x \leq y$ *we have* $i(x) \leq i\left(\frac{x+y}{2}\right) \leq i(y)$ *implies that*

$$i(y) e^{y-x} - i\left(\frac{x+y}{2}\right) e^{y-x} \geq 0 \geq i(x) - i\left(\frac{x+y}{2}\right).$$

Hence,

$$i(y) e^{y-x} - i(x) \geq i\left(\frac{x+y}{2}\right) (e^{y-x} - 1).$$

Multiplying the above inequality by $e^x/(y-x)$, *we get*

$$\frac{i(y) e^y - i(x) e^x}{y - x} \geq i\left(\frac{x+y}{2}\right) \frac{e^y - e^x}{y - x}.$$

Upon using Lemma 5, we obtain

$$\frac{i(y) e^y - i(x) e^x}{y - x} \geq i\left(\frac{x+y}{2}\right) \frac{e^y - e^x}{y - x} \geq i\left(\frac{x+y}{2}\right) e^{(x+y)/2}$$

which in turn

$$\frac{f(y) - f(x)}{y - x} \geq f\left(\frac{x+y}{2}\right).$$

Again for $y \leq x$, *we have* $i(y) \leq i\left(\frac{x+y}{2}\right) \leq i(x)$ *and we can deduce that*

$$i(y) - i\left(\frac{x+y}{2}\right) \leq 0 \leq i(x) e^{x-y} - i\left(\frac{x+y}{2}\right) e^{x-y}.$$

Hence,

$$i\left(\frac{x+y}{2}\right) (e^{x-y} - 1) \leq i(x) e^{x-y} - i(y).$$

Multiplying the last inequality by $e^y/(x-y)$ *we have*

$$i\left(\frac{x+y}{2}\right) \frac{e^x - e^y}{x - y} \leq \frac{i(x) e^x - i(y) e^y}{x - y},$$

that is,

$$i\left(\frac{x+y}{2}\right) \frac{e^y - e^x}{y - x} \leq \frac{i(y) e^y - i(x) e^x}{y - x}$$

implies that

$$i\left(\frac{x+y}{2}\right)\frac{e^y-e^x}{y-x}\leq\frac{f(y)-f(x)}{y-x}.$$

Using Lemma 5, it follows that

$$i\left(\frac{x+y}{2}\right)e^{\frac{x+y}{2}}\leq i\left(\frac{x+y}{2}\right)\frac{e^y-e^x}{y-x}\leq\frac{f(y)-f(x)}{y-x},$$

that is,

$$f\left(\frac{x+y}{2}\right)\leq\frac{f(y)-f(x)}{y-x}.$$

This completes the proof of the lemma.

Lemma 7 *If a function $f: I\to\mathbb{R}^+$ is nondecreasing and satisfies the inequality*

$$\frac{f(y)-f(x)}{y-x}\leq f\left(\frac{x+y}{2}\right) \tag{2.12}$$

for all $x<y$ in I, then f can be represented in the form $f(x)=d(x)e^x, x\in I$, with a nonincreasing function $d: I\to\mathbb{R}^+$.

Proof 35 *From* (2.12), *we get*

$$f(y)-f(x)\leq(y-x)f\left(\frac{x+y}{2}\right).$$

Put $x+h=y$ in last inequality. Then

$$f(x+h)\leq f(x)+hf\left(\frac{2x+h}{2}\right).$$

Therefore, for $h\in(0,1)$

$$0\leq f(x+h)-f(x)\leq hf\left(x+\frac{h}{2}\right)\leq hf(x+h)$$

implies that $(1-h)f(x+h)\leq f(x)$ and hence $f(x+h)\leq\frac{f(x)}{1-h}$. We claim that

$$f(x+ih)\leq\frac{f(x)}{(1-h)^i}$$

for $x+ih\in I$ and $i\in\mathbb{N}$. Our claim is true for $i=1$. Assume that it is true for $i=k$, that is, $f(x+kh)\leq\frac{f(x)}{(1-h)^k}$. For $i=k+1$,

$$f(x+(k+1)h)=f((x+h)+kh)\leq\frac{f(x+h)}{(1-h)^k}\leq\frac{f(x)}{(1-h)(1-h)^k}$$

implies that

$$f(x+(k+1)h) \leq \frac{f(x)}{(1-h)^{k+1}}.$$

So, our claim holds and it is true $\forall i \in \mathbb{N}$.

If $x \leq y$ *in* I *and* n *is sufficiently large, then* $\frac{(y-x)}{n} \in (0,1)$. *Hence we have*

$$f(y) = f\left(x+(y-x)\right) = f\left(x+n\frac{(y-x)}{n}\right) \leq \frac{f(x)}{\left(1-\frac{y-x}{n}\right)^n}.$$

When $n \to \infty$, $f(y) \leq f(x)e^{y-x}$, *that is,*

$$f(y)e^{-y} \leq f(x)e^{-x}, \forall x < y.$$

If, $d(x) = f(x)e^{-x}, x \in I$, *then it is a decreasing function. Hence the lemma is proved.*

Lemma 8 *A nondecreasing Jensen concave function* $f : I \to \mathbb{R}^+$ *satisfies the inequality (2.12) if and only if there exists a nonincreasing function* $d : I \to \mathbb{R}^+$ *such that* $I \ni x \mapsto d(x)e^x$ *is concave and* $f(x) = d(x)e^x, x \in I$.

Proof 36 *Necessary part is same as Lemma 7. To prove the sufficient part, assume that* $f(x) = d(x)e^x$ *for* $x \in I$ *is Jensen concave and nondecreasing with* d *nonincreasing. If* $x < y$ *in* I *then* $d(x) = f(x)e^{-x} \geq f(y)e^{-y} = d(y)$, *that is,*

$$f(x) \geq f(y)e^{x-y}.$$

Applying Lemma 5, we get

$$\frac{f(y)-f(x)}{y-x} \leq \frac{f(y)-f(y)e^{x-y}}{y-x} = \frac{f(y)}{e^y}\cdot\frac{e^y-e^x}{y-x} \leq \frac{f(y)}{e^y}\cdot\frac{e^x+e^y}{2}$$

$$= \frac{f(y)e^{x-y}+f(y)}{2} \leq \frac{f(x)+f(y)}{2} \leq f\left(\frac{x+y}{2}\right)$$

which is the inequality (2.12).

Theorem 18 *Given an* $\varepsilon > 0$ *let* $f : I \longrightarrow \mathbb{R}$ *be a differentiable function. Then*

$$|f'(x) - f(x)| \leq \varepsilon, x \in I$$

holds for all x *in* I *if and only if* f *can be represented in the form*

$$f(x) = \varepsilon + e^x l(e^{-x}), x \in I, \tag{2.13}$$

where l *is an arbitrary differentiable function defined on the interval*

$$J = \{e^{-x} | x \in I\}$$

nonincreasing and $2\varepsilon -$ *Lipschitz.*

Proof 37 *By* (2.1)

$$f(x) - \varepsilon \leq f'(x) \leq \varepsilon + f(x), \forall x \in I.$$

Let $g(x) = f(x) - \varepsilon$. *Then* $g(x) \leq g'(x)$ *and by Lemma 4(i), we can write*

$$f(x) - \varepsilon = i(x)e^x, x \in I,$$

that is,

$$f(x) = i(x)e^x + \varepsilon, x \in I, \tag{2.14}$$

where i *is differentiable and nondecreasing. On the other hand, if* $h(x) = f(x) + \varepsilon$ *with* $h^{'}(x) \leq h(x)$, *then we can write*

$$f(x) + \varepsilon = d(x)e^x, x \in I,$$

that is,

$$f(x) = d(x)e^x - \varepsilon, \tag{2.15}$$

where d is differentiable and nonincreasing. From (2.14) *and* (2.15), *it follows that*

$$i(x)e^x + \varepsilon = d(x)e^x - \varepsilon, x \in I. \tag{2.16}$$

Consequently,

$$i'(x)e^x + i(x)e^x = d'(x)e^x + e^x d(x) = d'(x)e^x + i(x)e^x + 2\varepsilon, x \in I.$$

Using the fact that d *is a nonincreasing function and so* $d' \leq 0$ *and*

$$d'(x) = \frac{i'(x)e^x - 2\varepsilon}{e^x} = i'(x) - 2\varepsilon e^{-x} \leq 0, x \in I$$

implies

$$0 \leq i'(x) \leq 2\varepsilon e^{-x}, x \in I.$$

Define $J = \{e^{-x} | x \in I\}$ *and* $l : J \to \mathbb{R}$ *be such that*

$$l(z) = i(-Inz), z \in J. \tag{2.17}$$

then l *is differentiable and*

$$l'(z) = -i'(-lnz)/z \leq 0, z \in J.$$

Hence l *is nonincreasing and also,* l *is* 2ε*-Lipschitz. Indeed* $\forall z_1, z_2$ *in* J, $z_1 \neq z_2$, *by the mean value theorem there exists* z_3 *in* $(min(z_1, z_2), max(z_1, z_2))$ *such that*

$$|l(z_1) - l(z_2)| = |l'(z_3)||z_1 - z_2| = \left| \frac{-i'(-\ln z_3)}{z_3} \right| |z_1 - z_2|$$

$$= i'(-\ln z_3)e^{-\ln z_3}|z_1 - z_2| \leq 2\varepsilon|z_1 - z_2|.$$

Thus by (2.17) *we have the required expression:*

$$f(x) = \varepsilon + i(x)e^x = \varepsilon + l(e^{-x})e^x.$$

Theorem 19 *Given an $\varepsilon > 0$, let $f : I \to \mathbb{R}^+$ be a function such that $f(x) \geq \varepsilon$ for all x in I. Then f satisfies the inequality*

$$f\left(\frac{x+y}{2}\right) \leq \frac{f(y)-f(x)}{y-x} + \varepsilon \tag{2.18}$$

for all $x < y$ in I, if and only if f can be represented in the form $f(x) = \varepsilon + i(x)e^x, x \in I$, where $i : I \to \mathbb{R}^+$ is a nondecreasing function.

Proof 38 *Applying the Lemma 6 to the function $f - \varepsilon$, we obtain a function $f : I \to \mathbb{R}^+$ satisfies*

$$f\left(\frac{x+y}{2}\right) - \varepsilon \leq \frac{(f(y)-\varepsilon)-(f(x)-\varepsilon)}{y-x},$$

for all $x \neq y$ in I, if and only if $f-\varepsilon$ can be represented in the form $f(x)-\varepsilon = i(x)e^x$, where $i : I \to \mathbb{R}^+$ is an arbitrary nondecreasing function. Hence

$$f\left(\frac{x+y}{2}\right) - \varepsilon \leq \frac{f(y)-f(x)}{y-x},$$

that is,

$$f\left(\frac{x+y}{2}\right) \leq \frac{f(y)-f(x)}{y-x} + \varepsilon$$

if and only if $f(x) = i(x)e^x + \varepsilon, x \in I$ which completes the proof of the theorem.

Theorem 20 *Given an $\varepsilon > 0$, a nondecreasing Jensen concave function $f : I \to \mathbb{R}$ satisfying $f(x) \geq -\varepsilon$ for all $x \in I$, is a solution of the inequality*

$$\frac{f(y)-f(x)}{y-x} - \varepsilon \leq f\left(\frac{x+y}{2}\right), \tag{2.19}$$

if and only if $f(x) = d(x)e^x - \varepsilon$, where $d : I \to \mathbb{R}^+$ is nonincreasing and $I \ni x \mapsto d(x)e^x$ is Jensen concave.

Proof 39 *Applying the Lemma 8 to the function $f + \varepsilon$, we obtain a function $f : I \to \mathbb{R}^+$ satisfying*

$$\frac{(f(y)+\varepsilon)-(f(x)+\varepsilon)}{y-x} \leq f\left(\frac{x+y}{2}\right) + \varepsilon$$

for all $x < y$ in I if and only if there exists a nonincreasing function $d : I \to \mathbb{R}^+$ such that $I \ni x \mapsto d(x)e^x$ is concave and $f(x)+\varepsilon = d(x)e^x, x \in I$. Hence,

$$\frac{f(y)-f(x)}{y-x} \leq f\left(\frac{x+y}{2}\right) + \varepsilon,$$

that is,

$$\frac{f(y)-f(x)}{y-x} - \varepsilon \leq f\left(\frac{x+y}{2}\right)$$

if and only if $f(x) = d(x)e^x - \varepsilon$, where $d : I \to \mathbb{R}^+$ is nonincreasing and $I \ni x \mapsto d(x)e^x$ is Jensen concave. Hence the theorem is proved.

2.2 Stability of $y'(t) = \lambda y(t)$

In this section, the Hyers-Ulam-Rassias stability of the following linear differential equation:

$$y'(t) = \lambda y(t) \tag{2.20}$$

is provided which generalizes a theorem of Alsina and Ger[1]. Moreover, it is proved that if λ is a non-zero real number and $I = (a, b)$ is an arbitrary open interval, and also if a continuously differentiable function $y : I \to \mathbb{R}$ satisfies the inequality

$$|y'(t) - \lambda y(t)| \leq \sum_{k=0}^{n} a_k t^k, \forall t \in I,$$

then there exist real numbers c and α_k, $k \in \{0, 1, ..., n\}$, such that

$$|y(t) - ce^{\lambda t}| \leq \begin{cases} |\sum_{k=0}^{n} \alpha_k t^k - \lim_{s\to b-} \sum_{k=0}^{n} \alpha_k s^k e^{\lambda(t-s)}| & (for \lambda > 0), \\ |\sum_{k=0}^{n} \alpha_k t^k - \lim_{s\to a+} \sum_{k=0}^{n} \alpha_k s^k e^{\lambda(t-s)}| & (for \lambda < 0), \end{cases}$$

for any $t \in I$.

Let $I = (a, b)$ be an open real interval, where $-\infty \leq a \leq b \leq \infty$. Suppose, real numbers $a_0, a_1, \cdots, a_n$ are given which satisfy the property:

$$\sum_{k=0}^{n} a_k t^k \geq 0 \;\; (t \in I), \tag{2.21}$$

where n is a fixed non-negative integer. Set

$$\alpha_k = \sum_{i=k}^{n} \frac{i!a_i}{k!\lambda^{i+1-k}} \;\; (k = 0, 1, \cdots, n) \tag{2.22}$$

and λ is true for some non-zero real number. Indeed,

$$\lambda\alpha_0 - \alpha_1 = \sum_{i=0}^{n} \frac{i!a_i}{\lambda^i} - \sum_{i=1}^{n} \frac{i!a_i}{\lambda^i} = a_0,$$

$$\lambda\alpha_1 - 2\alpha_2 = \sum_{i=1}^{n} \frac{i!a_i}{\lambda^{i-1}} - \sum_{i=2}^{n} \frac{i!a_i}{\lambda^{i-1}} = a_1$$

and hence

$$\lambda\alpha_{n-1} - n\alpha_n = \sum_{i=n-1}^{n} \frac{i!a_i}{(n-1)!\lambda^{i-n+1}} - \sum_{i=n}^{n} \frac{i!a_i}{(n-1)!\lambda^{i+1-n}} = a_{n-1}$$

implies that

$$\sum_{k=0}^{n}\lambda\alpha_k t^k - \sum_{k=1}^{n}k\alpha_k t^{k-1} = (\lambda\alpha_0 - \alpha_1) + (\lambda\alpha_1 - 2\alpha_2)t + \cdots + (\lambda\alpha_{n-1} - n\alpha_n)t^{n-1} + \lambda\alpha_n t^n = \sum_{k=0}^{n} a_k t^k, \quad (t \in \mathbb{R}). \tag{2.23}$$

Lemma 9 *Let I be an arbitrary non-degenerate open interval. Assume that $z : I \to \mathbb{R}$ is a continuously differentiable function and λ is a real number.*

(a) *The inequality $\lambda z(t) \leq z'(t)$ holds true for all $t \in I$ if and only if there exists an increasing continuously differentiable function $i : I \to \mathbb{R}$ such that*

$$z(t) = i(t)e^{\lambda t}$$

for all $t \in I$;

(b) *The inequality $\lambda z(t) \geq z'(t)$ holds true for any $t \in I$ if and only if there exists a decreasing continuously differentiable function $d : I \to \mathbb{R}$ such that*

$$z(t) = d(t)e^{\lambda t}$$

for all $t \in I$.

Proof 40 *The proof of the lemma follows from the proof of Lemma 4. Hence the details are omitted.*

Lemma 10 *For any non-zero real number λ and for any real sequence $\{a_i\}_{i=0}^{n}$, the equality*

$$\sum_{k=0}^{m}\sum_{i=k}^{m}\frac{i!a_i}{k!\lambda^{i+1-k}}t^k = \sum_{k=0}^{m}a_k\sum_{i=0}^{k}\frac{k!t^{k-i}}{(k-i)!\lambda^{i+1}}, (t \in \mathbb{R})$$

holds for any non-negative integer m.

Proof 41 *Our claim is true for $m = 0$. Let's assume that, it is true for $m = n \geq 0$, that is,*

$$\sum_{k=0}^{n}\sum_{i=k}^{n}\frac{i\,!a_i}{k\,!\lambda^{i+1-k}}t^k = \sum_{k=0}^{n}a_k\sum_{i=0}^{k}\frac{k\,!t^{k-i}}{(k-i)\,!\lambda^{i+1}}.$$

For $m = n+1$,

$$\begin{aligned}\sum_{k=0}^{n+1}\sum_{i=k}^{n+1}\frac{i\,!a_it^k}{k\,!\lambda^{i+1-k}} &= \sum_{k=0}^{n}\sum_{i=k}^{n+1}\frac{i\,!a_it^k}{k\,!\lambda^{i+1-k}} + \sum_{i=n+1}^{n+1}\frac{i\,!a_it^{n+1}}{(n+1)\,!\lambda^{i+1-(n+1)}}\\ &= \sum_{k=0}^{n}\sum_{i=k}^{n+1}\frac{i\,!a_it^k}{k\,!\lambda^{i+1-k}} + \frac{(n+1)\,!a_{n+1}t^{n+1}}{(n+1)\,!\lambda^{(n+2)-(n+1)}}\\ &= \sum_{k=0}^{n}\left(\sum_{i=k}^{n}\frac{i\,!a_it^k}{k\,!\lambda^{i+1-k}} + \frac{(n+1)\,!a_{n+1}t^k}{k\,!\lambda^{n+2-k}}\right) + \frac{(n+1)\,!a_{n+1}t^{n+1}}{(n+1)\,!\lambda}\\ &= \sum_{k=0}^{n}\sum_{i=k}^{n}\frac{i\,!a_it^k}{k\,!\lambda^{i+1-k}} + \left(\sum_{k=0}^{n}\frac{(n+1)\,!a_{n+1}t^k}{k\,!\lambda^{n+2-k}} + \frac{(n+1)\,!a_{n+1}t^{n+1}}{(n+1)\,!\lambda}\right)\\ &= \sum_{k=0}^{n}\sum_{i=k}^{n}\frac{i\,!a_i}{k\,!\lambda^{i+1-k}}t^k + \sum_{k=0}^{n+1}\frac{(n+1)!a_{n+1}t^k}{k\,!\lambda^{n+2-k}},\end{aligned}$$

that is,

$$\sum_{k=0}^{n+1}\sum_{i=k}^{n+1}\frac{i\,!a_it^k}{k\,!\lambda^{i+1-k}} = \sum_{k=0}^{n}a_k\sum_{i=0}^{k}\frac{k\,!t^{k-i}}{(k-i)\,!\lambda^{i+1}} + a_{n+1}\sum_{k=0}^{n+1}\frac{(n+1)!t^k}{k\,!\lambda^{n+2-k}}. \tag{2.24}$$

Putting $k = n+1-i$ *in the second sum of the equality* (2.24), *we obtain*

$$\begin{aligned}\sum_{k=0}^{n+1}\sum_{i=k}^{n+1}\frac{i!a_it^k}{k!\lambda^{i+1-k}} &= \sum_{k=0}^{n}a_k\sum_{i=0}^{k}\frac{k!t^{k-i}}{(k-i)!\lambda^{i+1}} + a_{n+1}\sum_{n+1-i=0}^{n+1-i=n+1}\frac{(n+1)!t^{n+1-i}}{(n+1-i)!\lambda^{i+1}}\\ &= \sum_{k=0}^{n}a_k\sum_{i=0}^{k}\frac{k!t^{k-i}}{(k-i)!\lambda^{i+1}} + a_{n+1}\sum_{i=n+1}^{i=0}\frac{(n+1)!t^{n+1-i}}{(n+1-i)!\lambda^{i+1}}\\ &= \sum_{k=0}^{n}a_k\sum_{i=0}^{k}\frac{k!t^{k-i}}{(k-i)!\lambda^{i+1}} + a_{n+1}\sum_{i=0}^{n+1}\frac{(n+1)!t^{n+1-i}}{(n+1-i)!\lambda^{i+1}}\\ &= \sum_{k=0}^{n+1}a_k\sum_{i=0}^{k}\frac{k!t^{k-i}}{(k-i)!\lambda^{i+1}}.\end{aligned}$$

So, our claim holds for any non-negative integer m. *Hence the lemma is proved.*

Theorem 21 *Let* I *be an arbitrary open interval and let* λ *be a non-zero real number. A continuously differentiable function* $y : I \to \mathbb{R}$ *satisfies the following inequality:*

$$|y'(t) - \lambda y(t)| \le \sum_{k=0}^{n} a_kt^k \tag{2.25}$$

for all $t \in I$ *if and only if there exists an increasing continuously differentiable function* $i : I \to \mathbb{R}$ *such that*

$$y(t) = i(t)e^{\lambda t} + \sum_{k=0}^{n} \alpha_k t^k \tag{2.26}$$

and

$$0 \leq i'(t) \leq 2\sum_{k=0}^{n} a_k t^k e^{-\lambda t} \tag{2.27}$$

for any $t \in I$.

Proof 42 *Let's assume that a continuously differentiable function* $y : I \to \mathbb{R}$, *satisfies* (2.25). *So,*

$$-\sum_{k=0}^{n} a_k t^k \leq y'(t) - \lambda y(t) \leq \sum_{k=0}^{n} a_k t^k,$$

that is,

$$\lambda y(t) - \sum_{k=0}^{n} a_k t^k \leq y'(t) \leq \lambda y(t) + \sum_{k=0}^{n} a_k t^k, \quad (\forall t \in I). \tag{2.28}$$

(2.28) *implies following two inequalities*

$$y'(t) - \lambda y(t) - \sum_{k=0}^{n} a_k t^k \leq 0 \tag{2.29}$$

and

$$y'(t) - \lambda y(t) + \sum_{k=0}^{n} a_k t^k \geq 0 \tag{2.30}$$

for each $t \in I$. *Let's define a function* $z_1 : I \to \mathbb{R}$ *such that*

$$z_1(t) = y(t) + \sum_{k=0}^{n} \alpha_k t^k.$$

Indeed, $z_1(t)$ *is continuously differentiable and*

$$z_1'(t) = y'(t) + \sum_{k=1}^{n} k\alpha_k t^{k-1}.$$

Therefore,

$$\begin{aligned} z_1'(t) - \lambda z_1(t) &= y'(t) + \sum_{k=1}^{n} k\alpha_k t^{k-1} - \lambda z_1(t) \\ &= y'(t) + \sum_{k=1}^{n} k\alpha_k t^{k-1} - \lambda\left(y(t) + \sum_{k=0}^{n} \alpha_k t^k\right) \\ &= y'(t) + \sum_{k=1}^{n} k\alpha_k t^{k-1} - \lambda y(t) - \lambda\sum_{k=0}^{n} \alpha_k t^k. \end{aligned} \tag{2.31}$$

Applying (2.23) *to* (2.31), *we get*

$$z_1'(t) - \lambda z_1(t) = (y'(t) - \lambda y(t)) - \sum_{k=0}^{n} a_k t^k$$

and hence (2.29) *becomes*

$$z_1'(t) - \lambda z_1(t) \leq 0, t \in I.$$

By Lemma 9(b), there exists a decreasing continuously differentiable function $d : I \to \mathbb{R}$ *such that*

$$z_1(t) = y(t) + \sum_{k=0}^{n} \alpha_k t^k = d(t)e^{\lambda t},$$

that is,

$$y(t) = d(t)e^{\lambda t} - \sum_{k=0}^{n} \alpha_k t^k, \quad \forall t \in I. \tag{2.32}$$

Consider, $z_2 : I \to \mathbb{R}$ *which is continuously differentiable function defined by*

$$z_2(t) = y(t) - \sum_{k=0}^{n} \alpha_k t^k.$$

Therefore,

$$z_2'(t) = y'(t) - \sum_{k=1}^{n} k\alpha_k t^{k-1}$$

implies that

$$\begin{aligned} z_2'(t) - \lambda z_2(t) &= y'(t) - \sum_{k=1}^{n} k\alpha_k t^{k-1} - \lambda \left(y(t) - \sum_{k=0}^{n} \alpha_k t^k \right), \\ &= (y'(t) - \lambda y(t)) + \left(\sum_{k=0}^{n} \lambda\alpha_k t^k - \sum_{k=1}^{n} k\alpha_k t^{k-1} \right). \end{aligned} \tag{2.33}$$

Applying (2.23), (2.33) *becomes*

$$z_2'(t) - \lambda z_2(t) = (y'(t) - \lambda y(t)) + \sum_{k=0}^{n} a_k t^k,$$

that is,

$$z_2'(t) - \lambda z_2(t) \geq 0, t \in I.$$

By Lemma 9(a), there exists an increasing continuously differentiable function $i : I \to \mathbb{R}$ *such that*

$$z_2(t) = y(t) - \sum_{k=0}^{n} \alpha_k t^k = i(t)e^{\lambda t}$$

implies that

$$y(t) = i(t)e^{\lambda t} + \sum_{k=0}^{n} \alpha_k t^k \tag{2.34}$$

which is (2.27)*. Comparing* (2.32) *and* (2.34)*, we obtain*

$$y(t) = d(t)e^{\lambda t} - \sum_{k=0}^{n} \alpha_k t^k = i(t)e^{\lambda t} + \sum_{k=0}^{n} \alpha_k t^k. \tag{2.35}$$

Differentiating (2.35) *with respect to* t*, we have*

$$d'(t)e^{\lambda t} + \lambda d(t)e^{\lambda t} - \sum_{k=1}^{n} k\alpha_k t^{k-1} = i'(t)e^{\lambda t} + \lambda i(t)e^{\lambda t} + \sum_{k=1}^{n} k\alpha_k t^{k-1},$$

that is,

$$\begin{aligned} &d'(t)e^{\lambda t} + \lambda \left(y(t) + \sum_{k=0}^{n} \alpha_k t^k \right) - \sum_{k=1}^{n} k\alpha_k t^{k-1} \\ &= i'(t)e^{\lambda t} + \lambda \left(y(t) - \sum_{k=0}^{n} \alpha_k t^k \right) + \sum_{k=1}^{n} k\alpha_k t^{k-1} \end{aligned}$$

due to (2.32) *and* (2.34)*. Consequently,*

$$d'(t)e^{\lambda t} + \sum_{k=0}^{n} \lambda\alpha_k t^k - \sum_{k=1}^{n} k\alpha_k t^{k-1} = i'(t)e^{\lambda t} - \sum_{k=0}^{n} \lambda\alpha_k t^k + \sum_{k=1}^{n} k\alpha_k t^{k-1},$$

implies that

$$d'(t)e^{\lambda t} = i'(t)e^{\lambda t} - 2\sum_{k=0}^{n} \lambda\alpha_k t^k + 2\sum_{k=1}^{n} k\alpha_k t^{k-1},$$

that is,

$$d'(t) = i'(t) + 2\left(\sum_{k=1}^{n} k\alpha_k t^{k-1} - \sum_{k=0}^{n} \lambda\alpha_k t^k \right) e^{-\lambda t}.$$

Since $d(t)$ *is decreasing, then* $d'(t) \leq 0$*. Hence,*

$$i'(t) - 2\left(\sum_{k=0}^{n} \lambda\alpha_k t^k - \sum_{k=1}^{n} k\alpha_k t^{k-1} \right) e^{-\lambda t} \leq 0.$$

By (2.23)*,*

$$i'(t) - 2\left(\sum_{k=0}^{n} a_k t^k \right) e^{-\lambda t} \leq 0.$$

Because $i(t)$ is an increasing function, so

$$0 \leq i'(t) \leq 2 \left(\sum_{k=0}^{n} a_k t^k \right) e^{-\lambda t}$$

which is the inequality (2.27).

Conversly, suppose that a continuously differentiable function $y : I \to \mathbb{R}$ is such that

$$y(t) = i(t)e^{-\lambda t} + \sum_{k=0}^{n} \alpha_k t^k,$$

where $i : I \to \mathbb{R}$ is a continuously differentiable function satisfying (2.27). *Now,*

$$y'(t) = i'(t)e^{\lambda t} + \lambda i(t)e^{\lambda t} + \sum_{k=1}^{n} k\alpha_k t^{k-1}, \quad \forall t \in I$$

implies that

$$\begin{aligned} y'(t) - \lambda y(t) &= i'(t)e^{\lambda t} + \lambda i(t)e^{\lambda t} + \sum_{k=1}^{n} k\alpha_k t^{k-1} - \lambda \left(i(t)e^{\lambda t} + \sum_{k=0}^{n} \alpha_k t^k \right), \\ &= i'(t)e^{\lambda t} - \left(\sum_{k=0}^{n} \lambda\alpha_k t^k - \sum_{k=1}^{n} k\alpha_k t^{k-1} \right). \end{aligned}$$

Upon using (2.23), *it follows that*

$$y'(t) - \lambda y(t) = i'(t)e^{\lambda t} - \sum_{k=0}^{n} a_k t^k. \tag{2.36}$$

By (2.27),

$$0 \leq i'(t)e^{\lambda t} \leq 2 \sum_{k=0}^{n} a_k t^k.$$

Ultimately,

$$-\sum_{k=0}^{n} a_k t^k \leq i'(t)e^{\lambda t} - \sum_{k=0}^{n} a_k t^k \leq \sum_{k=0}^{n} a_k t^k$$

and therefore,

$$-\sum_{k=0}^{n} a_k t^k \leq y'(t) - \lambda y(t) \leq \sum_{k=0}^{n} a_k t^k \forall t \in I$$

due to (2.36). *This completes the proof of the theorem.*

Theorem 22 *Let λ be a non-zero real constant. Assume that $I = (a, b)$ is an arbitrary open interval with $-\infty \leq a < b \leq \infty$. If a continuously differentiable function $y : I \to \mathbb{R}$ satisfies the inequality (2.25) for all $t \in I$, then there exists a real number c such that*

$$|y(t) - ce^{\lambda t}| \leq \begin{cases} |\sum_{k=0}^{n} \alpha_k t^k - \lim_{s\to b-} \sum_{k=0}^{n} \alpha_k s^k e^{\lambda(t-s)}| & (for \lambda > 0), \\ |\sum_{k=0}^{n} \alpha_k t^k - \lim_{s\to a+} \sum_{k=0}^{n} \alpha_k s^k e^{\lambda(t-s)}| & (for \lambda < 0), \end{cases} \tag{2.37}$$

for all $t \in I$.

Proof 43 *Consider the case when $\lambda > 0$. Define*

$$c = \lim_{s\to b-} \left\{ i(s) + \sum_{k=0}^{n} \alpha_k s^k e^{-\lambda s} \right\},$$

where $i : I \to \mathbb{R}$ is an increasing continuously differentiable function given in Theorem 21. Integrating (2.27) from t to b we obtain

$$0 \leq (i(\tau))_t^b \leq 2 \sum_{k=0}^{n} a_k \left(-\sum_{i=0}^{k} \frac{k! \tau^{k-i} e^{-\lambda\tau}}{(k-i)! \lambda^{i+1}} \right)_t^b, \tag{2.38}$$

where we have used the formula

$$\int \tau^k e^{-\lambda\tau} d\tau = -\sum_{i=0}^{k} \frac{k! \tau^{k-i}}{(k-i)! \lambda^{i+1}} e^{-\lambda\tau}$$

for $k = 0, 1, 2, \cdots$. Now, (2.38) implies that

$$0 \leq i(b) - i(t) \leq -2 \sum_{k=0}^{n} a_k \left(\sum_{i=0}^{k} \frac{k! \tau^{k-i} e^{-\lambda\tau}}{(k-i)! \lambda^{i+1}} \right)_t^b,$$

that is,

$$0 \leq \lim_{s\to b-} i(s) - i(t) \leq -2 \sum_{k=0}^{n} a_k \left(\sum_{i=0}^{k} \frac{k! b^{k-i} e^{-\lambda b}}{(k-i)! \lambda^{i+1}} - \sum_{i=0}^{k} \frac{k! t^{k-i} e^{-\lambda t}}{(k-i)! \lambda^{i+1}} \right)$$

$$\leq -2 \lim_{s\to b-} \sum_{k=0}^{n} a_k \sum_{i=0}^{k} \frac{k! s^{k-i} e^{-\lambda s}}{(k-i)! \lambda^{i+1}} + 2 \sum_{k=0}^{n} a_k \sum_{i=0}^{k} \frac{k! t^{k-i} e^{-\lambda t}}{(k-i)! \lambda^{i+1}}.$$

Applying Lemma 10, to the above inequality, we find

$$0 \leq \lim_{s\to b-} i(s) - i(t) \leq -2 \lim_{s\to b-} \left(\sum_{k=0}^{n} \sum_{i=k}^{n} \frac{i! a_i s^k}{k! \lambda^{i+1-k}} \right) e^{-\lambda s} + 2 \left(\sum_{k=0}^{n} \sum_{i=k}^{n} \frac{i! a_i t^k}{k! \lambda^{i+1-k}} \right) e^{-\lambda t},$$

that is, by using (2.22)

$$0 \le \lim_{s \to b-} i(s) - i(t) \le -2 \lim_{s \to b-} \sum_{k=0}^{n} \alpha_k s^k e^{-\lambda s} + 2 \sum_{k=0}^{n} \alpha_k t^k e^{-\lambda t}$$
$$\le c - \lim_{s \to b-} \sum_{k=0}^{n} \alpha_k s^k e^{-\lambda s} - i(t) \le -2 \lim_{s \to b-} \sum_{k=0}^{n} \alpha_k s^k e^{-\lambda s} + 2 \sum_{k=0}^{n} \alpha_k t^k e^{-\lambda t}. \tag{2.39}$$

Multiplying $e^{\lambda t}$ *to* (2.39), *we get*

$$0 \le ce^{\lambda t} - \lim_{s \to b-} \sum_{k=0}^{n} \alpha_k s^k e^{-\lambda s} e^{\lambda t} - i(t) e^{\lambda t} \le -2 \lim_{s \to b-} \sum_{k=0}^{n} \alpha_k s^k e^{-\lambda s} e^{\lambda t} + 2 \sum_{k=0}^{n} \alpha_k t^k e^{0},$$

that is, by adding $\lim_{s\to b-} \sum_{k=0}^{n} \alpha_k s^k e^{\lambda(t-s)} - \sum_{k=0}^{n} \alpha_k t^k$ *to the above inequality, we get*

$$\lim_{s \to b-} \sum_{k=0}^{n} \alpha_k s^k e^{\lambda(t-s)} - \sum_{k=0}^{n} \alpha_k t^k \le ce^{\lambda t} - \sum_{k=0}^{n} \alpha_k t^k - i(t) e^{\lambda t}$$
$$\le - \lim_{s \to b-} \sum_{k=0}^{n} \alpha_k s^k e^{\lambda(t-s)} + \sum_{k=0}^{n} \alpha_k t^k.$$

Therefore,

$$-\left(\sum_{k=0}^{n} \alpha_k t^k - \lim_{s \to b-} \sum_{k=0}^{n} \alpha_k s^k e^{\lambda(t-s)} \right) \le ce^{\lambda t} - \left(\sum_{k=0}^{n} \alpha_k t^k + i(t) e^{\lambda t} \right)$$
$$\le \sum_{k=0}^{n} \alpha_k t^k - \lim_{s \to b-} \sum_{k=0}^{n} \alpha_k s^k e^{\lambda(t-s)}. \tag{2.40}$$

By (2.26) *and* (2.40) *we obtain*

$$-\left(\sum_{k=0}^{n} \alpha_k t^k - \lim_{s \to b-} \sum_{k=0}^{n} \alpha_k s^k e^{\lambda(t-s)} \right) \le ce^{\lambda t} - y(t)$$
$$\le \sum_{k=0}^{n} \alpha_k t^k - \lim_{s \to b-} \sum_{k=0}^{n} \alpha_k s^k e^{\lambda(t-s)}.$$

Consequently,

$$|ce^{\lambda t} - y(t)| \le \sum_{k=0}^{n} \alpha_k t^k - \lim_{s \to b-} \sum_{k=0}^{n} \alpha_k s^k e^{\lambda(t-s)},$$

that is,

$$|y(t) - ce^{\lambda t}| \le \left| \sum_{k=0}^{n} \alpha_k t^k - \lim_{s \to b-} \sum_{k=0}^{n} \alpha_k s^k e^{\lambda(t-s)} \right|.$$

Let λ be a negative real constant and define

$$c = \lim_{s \to a+} \left\{ i(s) + \sum_{k=0}^{n} \alpha_k s^k e^{-\lambda s} \right\},$$

where i is the increasing continuously differentiable function defined by $i : I \to \mathbb{R}$. Now, integrating the inequality in (2.27) from a to t, we find

$$0 \leq (i(\tau))_a^t \leq 2 \sum_{k=0}^{n} a_k \left(- \sum_{i=0}^{k} \frac{k! \tau^{k-i} e^{-\lambda \tau}}{(k-i)! \lambda^{i+1}} \right)_a^t \tag{2.41}$$

implies that

$$0 \leq i(t) - i(a) \leq -2 \sum_{k=0}^{n} a_k \left(\sum_{i=0}^{k} \frac{k! \tau^{k-i} e^{-\lambda \tau}}{(k-i)! \lambda^{i+1}} \right)_a^t,$$

that is,

$$\begin{aligned} 0 \leq i(t) - \lim_{s \to a+} i(s) &\leq -2 \sum_{k=0}^{n} a_k \left(\sum_{i=0}^{k} \frac{k! t^{k-i} e^{-\lambda t}}{(k-i)! \lambda^{i+1}} - \sum_{i=0}^{k} \frac{k! a^{k-i} e^{-\lambda a}}{(k-i)! \lambda^{i+1}} \right) \\ &\leq -2 \sum_{k=0}^{n} a_k \sum_{i=0}^{k} \frac{k! t^{k-i} e^{-\lambda t}}{(k-i)! \lambda^{i+1}} + 2 \lim_{s \to a+} \sum_{k=0}^{n} a_k \sum_{i=0}^{k} \frac{k! s^{k-i} e^{-\lambda s}}{(k-i)! \lambda^{i+1}}. \end{aligned}$$

Applying Lemma 10 to the above inequality, we find

$$0 \leq i(t) - \lim_{s \to a+} i(s) \leq -2 \left(\sum_{k=0}^{n} \sum_{i=k}^{n} \frac{i! a_i t^k}{k! \lambda^{i+1-k}} \right) e^{-\lambda t} + 2 \lim_{s \to a+} \left(\sum_{k=0}^{n} \sum_{i=k}^{n} \frac{i! a_i s^k}{k! \lambda^{i+1-k}} \right) e^{-\lambda s},$$

that is, by using (2.22)

$$\begin{aligned} 0 \leq i(t) - \lim_{s \to a+} i(s) &\leq -2 \sum_{k=0}^{n} \alpha_k t^k e^{-\lambda t} + 2 \lim_{s \to a+} \sum_{k=0}^{n} \alpha_k s^k e^{-\lambda s} \\ \leq i(t) - c + \lim_{s \to a+} \sum_{k=0}^{n} \alpha_k s^k e^{-\lambda s} &\leq -2 \sum_{k=0}^{n} \alpha_k t^k e^{-\lambda t} + 2 \lim_{s \to a+} \sum_{k=0}^{n} \alpha_k s^k e^{-\lambda s}. \end{aligned} \tag{2.42}$$

Multiplying $e^{\lambda t}$ to (2.42), we get

$$0 \leq i(t) e^{\lambda t} - c e^{\lambda t} + \lim_{s \to a+} \sum_{k=0}^{n} \alpha_k s^k e^{-\lambda s} e^{\lambda t} \leq -2 \sum_{k=0}^{n} \alpha_k t^k e^0 + 2 \lim_{s \to a+} \sum_{k=0}^{n} \alpha_k s^k e^{-\lambda s} e^{\lambda t},$$

that is, by adding $\sum_{k=0}^{n} \alpha_k t^k - \lim_{s \to a+} \sum_{k=0}^{n} \alpha_k s^k e^{\lambda(t-s)}$ to the above inequality, we get

$$\begin{aligned} \sum_{k=0}^{n} \alpha_k t^k - \lim_{s \to a+} \sum_{k=0}^{n} \alpha_k s^k e^{\lambda(t-s)} &\leq i(t) e^{\lambda t} - c e^{\lambda t} + \sum_{k=0}^{n} \alpha_k t^k \\ &\leq -\sum_{k=0}^{n} \alpha_k t^k + \lim_{s \to a+} \sum_{k=0}^{n} \alpha_k s^k e^{\lambda(t-s)}. \end{aligned}$$

Therefore,

$$-\left(\lim_{s\to a+}\sum_{k=0}^{n}\alpha_k s^k e^{\lambda(t-s)}-\sum_{k=0}^{n}\alpha_k t^k\right)\leq\left(i(t)e^{\lambda t}+\sum_{k=0}^{n}\alpha_k t^k\right)-ce^{\lambda t}$$
$$\leq\lim_{s\to a+}\sum_{k=0}^{n}\alpha_k s^k e^{\lambda(t-s)}-\sum_{k=0}^{n}\alpha_k t^k. \tag{2.43}$$

By (2.26) *and* (2.43) *we obtain*

$$-\left(\lim_{s\to a+}\sum_{k=0}^{n}\alpha_k s^k e^{\lambda(t-s)}-\sum_{k=0}^{n}\alpha_k t^k\right)\leq y(t)-ce^{\lambda t}$$
$$\leq\lim_{s\to a+}\sum_{k=0}^{n}\alpha_k s^k e^{\lambda(t-s)}-\sum_{k=0}^{n}\alpha_k t^k.$$

Consequently,

$$|y(t)-ce^{\lambda t}|\leq\lim_{s\to a+}\sum_{k=0}^{n}\alpha_k s^k e^{\lambda(t-s)}-\sum_{k=0}^{n}\alpha_k t^k,$$

that is,

$$|y(t)-ce^{\lambda t}|\leq\left|\sum_{k=0}^{n}\alpha_k t^k-\lim_{s\to a+}\sum_{k=0}^{n}\alpha_k s^k e^{\lambda(t-s)}\right|.$$

Hence, the theorem is proved.

Corollary 15 *Let λ be a non-zero real constant and define*

$$I=\begin{cases}(a,\infty) & (for\quad \lambda>0),\\ (-\infty,b) & (for\quad \lambda<0),\end{cases}$$

where $a\in\mathbb{R}\cup\{-\infty\}$ and $b\in\mathbb{R}\cup\{\infty\}$ are fixed. If a continuously differentiable function $y:I\to\mathbb{R}$ satisfies the inequality (2.25) *for all $t\in I$, then there exists a unique real number c such that*

$$|y(t)-ce^{\lambda t}|\leq\left|\sum_{k=0}^{n}\alpha_k t^k\right| \tag{2.44}$$

for all $t\in I$.

Proof 44 *According to Theorem 22, there exists a constant $c\in\mathbb{R}$ such that the inequality* (2.44) *holds for all $t\in I$. Hence, we only prove the uniqueness* *of c. Let c_1 is another real number with which the inequality (2.44) is satisfied. Now,*

$$
\begin{aligned}
|c - c_1| &= |ce^{-\lambda t}e^{\lambda t} - c_1 e^{-\lambda t}e^{\lambda t} + e^{-\lambda t}y(t) - e^{-\lambda t}y(t)| \\
&= |e^{-\lambda t}y(t) - c_1 e^{-\lambda t}e^{\lambda t} + ce^{-\lambda t}e^{\lambda t} - e^{-\lambda t}y(t)| \\
&\leq e^{-\lambda t}|y(t) - c_1 e^{-\lambda t}| + e^{-\lambda t}|ce^{\lambda t} - y(t)| \\
&= e^{-\lambda t}|y(t) - c_1 e^{-\lambda t}| + e^{-\lambda t}|y(t) - ce^{\lambda t}| \\
&\leq 2e^{-\lambda t}\left|\sum_{k=0}^{n} \alpha_k t^k\right| \\
&\to 0,
\end{aligned}
$$

when $t \to \infty$ for $\lambda > 0$ or $t \to -\infty$ for $\lambda < 0$, which completes the proof.

Theorem 23 *Let a be a real constant and let $I = (a, b)$ be an arbitrary non-degenerate open interval. Assume that a function $y : I \cup \{a\} \to \mathbb{R}$ is continuously differentiable on I and continuous at a on the right. If y satisfies the inequality (2.25) for any $t \in I$, then*

$$
\begin{aligned}
&\left(y(a) - \sum_{k=0}^{n} \alpha_k a^k\right) e^{\lambda(t-a)} + \sum_{k=0}^{n} \alpha_k t^k \leq y(t) \\
&\qquad \leq \left(y(a) + \sum_{k=0}^{n} \alpha_k a^k\right) e^{\lambda(t-a)} - \sum_{k=0}^{n} \alpha_k t^k
\end{aligned}
$$

for all $t \in I$.

Proof 45 *Let $c = \lim_{s \to a+} i(s)$, where $i : I \to \mathbb{R}$ is from Theorem 21. Integrating (2.27) from a to t, we obtain*

$$
0 \leq (i(\tau))_a^t \leq 2\sum_{k=0}^{n} a_k \left(-\sum_{i=0}^{k} \frac{k!\tau^{k-i}}{(k-i)!\lambda^{i+1}} e^{-\lambda\tau}\right)_a^t,
$$

that is,

$$
0 \leq i(t) - i(a) \leq -2\sum_{k=0}^{n} a_k \sum_{i=0}^{k} \frac{k!t^{k-i}e^{-\lambda t}}{(k-i)\,!\lambda^{i+1}} + 2\sum_{k=0}^{n} a_k \sum_{i=0}^{k} \frac{k\,!a^{k-i}e^{-\lambda a}}{(k-i)\,!\lambda^{i+1}}.
$$

Consequently,

$$
0 \leq i(t) - c \leq -2\sum_{k=0}^{n}\sum_{i=k}^{n} \frac{i!a_i t^k e^{-\lambda t}}{k!\lambda^{i+1-k}} + 2\sum_{k=0}^{n}\sum_{i=k}^{n} \frac{i!a_i a^k e^{-\lambda a}}{k!\lambda^{i+1-k}}
$$

due to Lemma 10. Hence,

$$
0 \leq i(t) - c \leq -2\sum_{k=0}^{n} \alpha_k t^k e^{-\lambda t} + 2\sum_{k=0}^{n} \alpha_k a^k e^{-\lambda a}
$$

due to (2.22). *Thus,*

$$0 \le i(t)e^{\lambda t} - ce^{\lambda t} \le -2\sum_{k=0}^{n} \alpha_k t^k + 2\sum_{k=0}^{n} \alpha_k a^k e^{\lambda(t-a)}$$

implies that

$$\sum_{k=0}^{n} \alpha_k t^k \le i(t)e^{\lambda t} + \sum_{k=0}^{n} \alpha_k t^k - ce^{\lambda t} \le -\sum_{k=0}^{n} \alpha_k t^k + 2\sum_{k=0}^{n} \alpha_k a^k e^{\lambda(t-a)}.$$

Upon using (2.26), *the above inequality yields*

$$\sum_{k=0}^{n} \alpha_k t^k \le y(t) - ce^{\lambda t} \le -\sum_{k=0}^{n} \alpha_k t^k + 2\sum_{k=0}^{n} \alpha_k a^k e^{\lambda(t-a)},$$

that is,

$$ce^{\lambda t} + \sum_{k=0}^{n} \alpha_k t^k \le y(t) \le ce^{\lambda t} + 2\sum_{k=0}^{n} \alpha_k a^k e^{\lambda(t-a)} - \sum_{k=0}^{n} \alpha_k t^k$$

implies that

$$ce^{\lambda t} + \sum_{k=0}^{n} \alpha_k t^k \le y(t) \le \left(ce^{\lambda a} + 2\sum_{k=0}^{n} \alpha_k a^k \right) e^{\lambda(t-a)} - \sum_{k=0}^{n} \alpha_k t^k. \quad (2.45)$$

Taking $t \to a+$ *in* (2.26), *we obtain*

$$y(a) = i(a)e^{\lambda a} + \sum_{k=0}^{n} \alpha_k a^k = ce^{\lambda a} + \sum_{k=0}^{n} \alpha_k a^k. \quad (2.46)$$

As $t \to a+$, (2.45) *becomes*

$$ce^{\lambda a} + \sum_{k=0}^{n} \alpha_k a^k \le y(a) \le ce^{\lambda a} + 2\sum_{k=0}^{n} \alpha_k a^k - \sum_{k=0}^{n} \alpha_k a^k,$$

that is,

$$y(a) = ce^{\lambda a} + \sum_{k=0}^{n} \alpha_k a^k$$

which is same as (2.46). *Hence, the theorem is proved.*

Example 2 *Let* $y : I \to \mathbb{R}$ *be a continuously differentiable function that satisfy* (2.25). *Choose* $a_0 = \varepsilon$ *and* $a_k = 0$, *for* $k \in \mathbb{N}$. *Then*

$$|y'(t) - \lambda y(t)| \le \varepsilon, \ \ \forall t \in I.$$

By (2.22), $\alpha_0 = \varepsilon\lambda^{-1}$ *and* $\alpha_k = 0$, *for* $k \in \mathbb{N}$. *By Corollary 15, there exists a unique real number* c *such that*

$$|y(t) - ce^{\lambda t}| \le \varepsilon|\lambda|^{-1}, \ \ \forall t \in I,$$

which is the Hyers-Ulam stability of the differential equation (2.20).

Example 3 *Assume that a_0 and a_1 satisfy $a_0 + a_1 t \geq 0, \ \forall t \in I$. By (2.22), $\alpha_0 = a_0\lambda^{-1} + a_1\lambda^{-2}$ and $\alpha_1 = a_1\lambda^{-1}$ and $a_k = 0, k = 2, 3, 4, \ldots$. Let a continuously differentiable function $y : I \to \mathbb{R}$ be satisfy (2.25), that is,*

$$|y'(t) - \lambda y(t)| \leq a_0 + a_1 t \ \ \forall t \in I.$$

Then by Corollary 15, there exists a unique real number c such that

$$|y(t) - ce^{\lambda t}| \leq |a_0\lambda^{-1} + a_1\lambda^{-2} + a_1\lambda^{-1} t| \forall t \in I,$$

which is the Hyers-Ulam stability of the differential equation (2.20).

2.3 Stability of $\varphi(t)y'(t) = y(t)$

This section deals with the Hyers-Ulam-Rassias stability of the following linear differential equation:

$$\varphi(t)y'(t) = y(t), \ \ t \in I, \tag{2.47}$$

where $I = (a, b)$ is an arbitrary open interval and we assume that a and b satisfy $-\infty \leq a < b < +\infty$. Also, we assume that $\varphi : I \to \mathbb{R}$ is a given function for which $\int_a^t d\tau/\varphi(\tau)$ exists for any $t \in I$. In brief, we prove that if either $\varphi(t) \geq 0$ satisfies $\forall t \in I$ or $\varphi(t) < 0$ holds $\forall t \in I$, and also if a differentiable function $y : I \to \mathbb{R}$ satisfies the inequality $|\varphi(t)y'(t) - y(t)| \leq \varepsilon$ for all $t \in I$, then there exists a real number c such that

$$\left| y(t) - c \ \exp\left\{ \int_a^t \frac{d\tau}{\varphi(\tau)} \right\} \right| \leq \varepsilon, \ \ \forall t \in I.$$

Remark 2 *If $\varphi(t) = \frac{1}{\lambda}$, $\lambda \in \mathbb{R} \setminus \{0\}$, $t \in I$, then (2.47) becomes (2.20). Otherwise, (2.47) is more general than (2.20). Hence, the results followed by (2.47) are more general results than (2.20).*

Lemma 11 *Suppose a differentiable function $z : I \to \mathbb{R}$ is given.*

(a) *The inequality $z(t) \leq \varphi(t)z'(t)$ is true for all $t \in I$ if and only if there exists a differentiable function $\alpha : I \to \mathbb{R}$ such that $\alpha'(t)\varphi(t) \geq 0$ and*

$$z(t) = \alpha(t) \exp\left\{ \int_a^t \frac{d\tau}{\varphi(\tau)} \right\},$$

for all $t \in I$.

(*b*) *The inequality* $z(t) \geq \varphi(t)z'(t)$ *holds true for any* $t \in I$ *if and only if there exists a differentiable function* $\beta : I \to \mathbb{R}$ *such that* $\beta'(t)\varphi(t) \leq 0$ *and*

$$z(t) = \beta(t)\exp\left\{\int_a^t \frac{d\tau}{\varphi(\tau)}\right\}$$

for all $t \in I$.

Proof 46 (*a*) *Suppose that* $z(t) \leq \varphi(t)z'(t)$ *is true for all* $t \in I$. *To prove* $\alpha'(t)\varphi(t) \geq 0, \ \forall t \in I$. *Define a function* $\alpha : I \to \mathbb{R}$ *such that*

$$\alpha(t) = \exp\left\{-\int_a^t \frac{d\tau}{\varphi(\tau)}\right\}z(t).$$

Then,

$$\alpha'(t) = \frac{1}{\varphi(t)}\exp\left\{-\int_a^t \frac{d\tau}{\varphi(\tau)}\right\}(\varphi(t)z'(t) - z(t)).$$

So, $\alpha'(t)\varphi(t) \geq 0$ *which is true for all* $t \in I$.

Conversely, let there exist a differentiable function $\alpha : I \to \mathbb{R}$ *with* $\alpha'(t)\varphi(t) \geq 0$, *for each* $t \in I$. *To prove that* $z(t) \leq \varphi(t)z'(t)$ *is true for all* $t \in I$. *If*

$$z(t) = \alpha(t)\exp\left\{\int_a^t \frac{d\tau}{\varphi(\tau)}\right\},$$

then

$$z'(t) = \alpha'(t)\exp\left\{\int_a^t \frac{d\tau}{\varphi(\tau)}\right\} + \frac{z(t)}{\varphi(t)},$$

that is,

$$\varphi(t)z'(t) = \alpha'(t)\varphi(t)\exp\left\{\int_a^t \frac{d\tau}{\varphi(\tau)}\right\} + z(t) \geq z(t)$$

for all $t \in I$.

(*b*) *Suppose that* $z(t) \geq \varphi(t)z'(t)$ *is true for all* $t \in I$. *To prove* $\beta'(t)\varphi(t) \leq 0 \ \forall t \in I$. *Define a function* $\beta : I \to \mathbb{R}$ *such that*

$$\beta(t) = \exp\left\{-\int_a^t \frac{d\tau}{\varphi(\tau)}\right\}z(t).$$

Then

$$\beta'(t) = \frac{1}{\varphi(t)}\exp\left\{-\int_a^t \frac{d\tau}{\varphi(\tau)}\right\}(\varphi(t)z'(t) - z(t)),$$

So, $\beta'(t)\varphi(t) \leq 0$ *which is true for all* $t \in I$.

Conversely, assume that there exists a differentiable function $\beta : I \to \mathbb{R}$ *with* $\beta'(t)\varphi(t) \leq 0$, *for each* $t \in I$. *To prove* $z(t) \geq \varphi(t)z'(t)$ *is true for all* $t \in I$. *If*

$$z(t) = \beta(t) \exp\left\{\int_a^t \frac{d\tau}{\varphi(\tau)}\right\},$$

then,

$$z'(t) = \beta'(t) \exp\left\{\int_a^t \frac{d\tau}{\varphi(\tau)}\right\} + \frac{z(t)}{\varphi(t)},$$

that is,

$$\varphi(t)z'(t) = \beta'(t)\varphi(t) \exp\left\{\int_a^t \frac{d\tau}{\varphi(\tau)}\right\} + z(t) \leq z(t)$$

for all $t \in I$. *This completes the proof of the lemma.*

Theorem 24 *Given an* $\varepsilon > 0$, *a differentiable function* $y : I \to \mathbb{R}$ *is a solution of the following inequality:*

$$|\varphi(t)y'(t) - y(t)| \leq \varepsilon \tag{2.48}$$

for all $t \in I$ *if and only if there exists a differentiable function* $\alpha : I \to \mathbb{R}$ *such that*

$$y(t) = \varepsilon + \alpha(t) \exp\left\{\int_a^t \frac{d\tau}{\varphi(\tau)}\right\} \tag{2.49}$$

and

$$0 \leq \alpha'(t)\varphi(t) \leq 2\varepsilon \exp\left\{-\int_a^t \frac{d\tau}{\varphi(\tau)}\right\} \tag{2.50}$$

for any $t \in I$.

Proof 47 *Suppose that a differentiable function* $y : I \to \mathbb{R}$ *is a solution of* (2.48), *that is,* y *satisfies the inequality*

$$y(t) - \varepsilon \leq \varphi(t)y'(t) \leq y(t) + \varepsilon \tag{2.51}$$

for all $t \in I$. *To prove that there exists a differentiable function* $\alpha : I \to \mathbb{R}$ *such that the function* $\alpha(t)$ *satisfies* (2.49) *and* (2.50). *Define* $z_1(t) = y(t) - \varepsilon$. *Hence from* (2.51), $z_1(t) \leq \varphi(t)z_1'(t), \ \forall t \in I$. *By Lemma 11(a), there exists a differentiable function* $\alpha : I \to \mathbb{R}$ *such that*

$$z_1(t) = \alpha(t) \exp\left\{\int_a^t \frac{d\tau}{\varphi(\tau)}\right\},$$

that is,

$$y(t) - \varepsilon = \alpha(t) \exp\left\{\int_a^t \frac{d\tau}{\varphi(\tau)}\right\}.$$

So, (2.49) *holds for all* $t \in I$, *where* α *satisfies* $\alpha'(t)\varphi(t) \geq 0$ *for all* $t \in I$ *due to Lemma 11(a). If we choose,* $z_2(t) = y(t) + \varepsilon$, *then* (2.51) *implies that* $z_2(t) \geq \varphi(t)z_2'(t), \quad \forall t \in I$. *By Lemma 11(b), there exists a differentiable function* $\beta : I \to \mathbb{R}$ *such that*

$$z_2(t) = \beta(t) \exp\left\{\int_a^t \frac{d\tau}{\varphi(\tau)}\right\},$$

that is,

$$y(t) + \varepsilon = \beta(t) \exp\left\{\int_a^t \frac{d\tau}{\varphi(\tau)}\right\} \tag{2.52}$$

and $\beta'(t)\varphi(t) \leq 0, \quad \forall t \in I$ *due to Lemma 11(b). Using* (2.49) *and* (2.52), *we obtain*

$$y'(t) = \alpha'(t) \exp\left\{\int_a^t \frac{d\tau}{\varphi(\tau)}\right\} + \frac{\alpha(t)}{\varphi(t)} \exp\left\{\int_a^t \frac{d\tau}{\varphi(\tau)}\right\} \tag{2.53}$$

and

$$\begin{aligned} y'(t) &= \beta'(t) \exp\left\{\int_a^t \frac{d\tau}{\varphi(\tau)}\right\} + \frac{\beta(t)}{\varphi(t)} \left(\exp\left\{\int_a^t \frac{d\tau}{\varphi(\tau)}\right\}\right) \\ &= \beta'(t) \exp\left\{\int_a^t \frac{d\tau}{\varphi(\tau)}\right\} + \frac{1}{\varphi(t)} (y(t) + \varepsilon) \end{aligned}$$

respectively. Again, we use (2.49) *in the last relation to get*

$$y'(t) = \beta'(t) \exp\left\{\int_a^t \frac{d\tau}{\varphi(\tau)}\right\} + \frac{1}{\varphi(t)} \left(2\varepsilon + \alpha(t) \exp\left\{\int_a^t \frac{d\tau}{\varphi(\tau)}\right\}\right). \tag{2.54}$$

Comparing (2.53) *and* (2.54), *it follows that*

$$\alpha'(t) \exp\left\{\int_a^t \frac{d\tau}{\varphi(\tau)}\right\} = \beta'(t) \exp\left\{\int_a^t \frac{d\tau}{\varphi(\tau)}\right\} + \frac{2\varepsilon}{\varphi(t)},$$

that is,

$$0 \geq \beta'(t)\varphi(t) = \alpha'(t)\varphi(t) - 2\varepsilon \exp\left\{-\int_a^t \frac{d\tau}{\varphi(\tau)}\right\}$$

implies that

$$0 \leq \alpha'(t)\varphi(t) \leq 2\varepsilon \exp\left\{-\int_a^t \frac{d\tau}{\varphi(\tau)}\right\}.$$

Conversely, assume that there exists a differentiable function $\alpha : I \to \mathbb{R}$ *such that the function* $\alpha(t)$ *satisfies* (2.49) *and* (2.50). *To prove that a* *differentiable function $y : I \to \mathbb{R}$ is a solution of (2.48). Clearly, (2.53) holds, that is,*

$$y'(t)\varphi(t) = \alpha'(t)\varphi(t)\exp\left\{\int_a^t \frac{d\tau}{\varphi(\tau)}\right\} + \alpha(t)\exp\left\{\int_a^t \frac{d\tau}{\varphi(\tau)}\right\}.$$

Consequently,

$$y'(t)\varphi(t) - y(t) = \alpha'(t)\varphi(t)\exp\left\{\int_a^t \frac{d\tau}{\varphi(\tau)}\right\} - \varepsilon$$

due to (2.49), that is,

$$\begin{aligned}(y'(t)\varphi(t) - y(t) + \varepsilon)\exp\left\{-\int_a^t \frac{d\tau}{\varphi(\tau)}\right\} &= \alpha'(t)\varphi(t) \\ &\leq 2\varepsilon\exp\left\{-\int_a^t \frac{d\tau}{\varphi(\tau)}\right\}\end{aligned}$$

due to (2.50). Therefore, $0 \leq (y'(t)\varphi(t) - y(t) + \varepsilon) \leq 2\varepsilon$ and $|\varphi(t)y'(t) - y(t)| \leq \varepsilon$. Hence, the theorem is proved.

Theorem 25 *If either $\varphi(t) > 0$ holds for all $t \in I$ or $\varphi(t) < 0$ holds for all $t \in I$, and if a differentiable function $y : I \to \mathbb{R}$ satisfies inequality (2.48) for all $t \in I$, then there exists a real number c such that*

$$\left|y(t) - c\exp\left\{\int_a^t \frac{d\tau}{\varphi(\tau)}\right\}\right| \leq \varepsilon \tag{2.55}$$

for any $t \in I$.

Proof 48 *Suppose that $\varphi(t) > 0$, for all $t \in I$ and consider a differentiable function $y : I \to \mathbb{R}$ which satisfies (2.48) $\forall\ t \in I$. Define $c = \lim_{t\to b-}\alpha(t)$, where $\alpha : I \to \mathbb{R}$. Using the fact that $\varphi'(t) > 0$, (2.50) becomes*

$$0 \geq -\alpha'(t) \geq -\frac{2\varepsilon}{\varphi(t)}\exp\left\{-\int_a^t \frac{d\tau}{\varphi(\tau)}\right\}.$$

Integrating the above inequality from t to b, we obtain

$$0 \geq -\left(\alpha(s)\right)_t^b \geq 2\varepsilon\int_t^b \left(\frac{-1}{\varphi(s)}\right)\exp\left\{-\int_a^s \frac{d\tau}{\varphi(\tau)}\right\}ds,$$

that is,

$$0 \geq \alpha(t) - \alpha(b) \geq 2\varepsilon\left(\exp\left\{-\int_a^s \frac{d\tau}{\varphi(\tau)}\right\}\right)_t^b.$$

Thus,

$$0 \geq \alpha(t) - c \geq 2\varepsilon \exp\left\{ - \int_a^b \frac{d\tau}{\varphi(\tau)} \right\} - 2\varepsilon \exp\left\{ - \int_a^t \frac{d\tau}{\varphi(\tau)} \right\}$$

implies that

$$0 \geq (\alpha(t) - c) \exp\left\{ \int_a^t \frac{d\tau}{\varphi(\tau)} \right\} \geq 2\varepsilon \exp\left\{ \int_b^t \frac{d\tau}{\varphi(\tau)} \right\} - 2\varepsilon,$$

that is,

$$\varepsilon \geq (\alpha(t) - c) \exp\left\{ \int_a^t \frac{d\tau}{\varphi(\tau)} \right\} + \varepsilon \geq 2\varepsilon \exp\left\{ - \int_t^b \frac{d\tau}{\varphi(\tau)} \right\} - \varepsilon \geq -\varepsilon.$$

Consequently, the above inequality becomes

$$\begin{aligned} \varepsilon &\geq \varepsilon + \alpha(t) \exp\left\{ \int_a^t \frac{d\tau}{\varphi(\tau)} \right\} - c \exp\left\{ \int_a^t \frac{d\tau}{\varphi(\tau)} \right\} \\ &\geq 2\varepsilon \exp\left\{ - \int_t^b \frac{d\tau}{\varphi(\tau)} \right\} - \varepsilon \geq - \varepsilon \end{aligned}$$

which is because of (2.49) *and equivalent to*

$$\varepsilon \geq y(t) - c \exp\left\{ \int_a^t \frac{d\tau}{\varphi(\tau)} \right\} \geq -\varepsilon.$$

Next, we assume that $\varphi(t) < 0$ *and consider* $c = \lim_{t \to a+} \alpha(t)$. *Dividing the inequalities* (2.50) *by* $\varphi(t)$, *we get*

$$0 \geq \alpha'(t) \geq \frac{2\varepsilon}{\varphi(t)} \exp\left\{ - \int_a^t \frac{d\tau}{\varphi(\tau)} \right\}.$$

Integrating the above inequality from a *to* t, *we obtain*

$$0 \geq (\alpha'(t))_a^t \geq 2\varepsilon \int_a^t \frac{1}{\varphi(s)} \exp\left\{ - \int_a^s \frac{d\tau}{\varphi(\tau)} \right\} ds,$$

that is,

$$0 \geq \alpha(t) - \alpha(a) \geq -2\varepsilon \left(\exp\left\{ - \int_a^s \frac{d\tau}{\varphi(\tau)} \right\} \right)_a^t.$$

The rest of the proof follows from the above case. This completes the proof of the theorem.

Corollary 16 *Assume that a is a real number and that $\varphi(t) > 0$ holds for all $t \in I$. Let a function $y : I \cup \{a\} \to \mathbb{R}$ be differentiable on I and continuous at a on the right. If y satisfies (2.48) for all $t \in I$ and for some $\varepsilon > O$, then*

$$(y(a)-\varepsilon)\exp\left\{\int_a^t \frac{d\tau}{\varphi(\tau)}\right\}+\varepsilon \le y(t) \le (y(a)+\varepsilon)\exp\left\{\int_a^t \frac{d\tau}{\varphi(\tau)}\right\}-\varepsilon \quad (2.56)$$

for any $t \in I$.

Proof 49 *If y satisfies (2.48) for any $t \in I$, then Theorem 24 implies that there exists a differentiable function $\alpha : I \to \mathbb{R}$ satisfying (2.49) and (2.50) for all $t \in I$. Define $c = \lim_{t \to a+} \alpha(t)$, where $\alpha : I \to \mathbb{R}$ is given in Theorem 24. Since $\varphi(t) > 0$ for $t \in I$, (2.50) becomes*

$$0 \le \alpha'(t) \le \frac{2\varepsilon}{\varphi(t)} \exp\left\{-\int_a^t \frac{d\tau}{\varphi(\tau)}\right\}. \quad (2.57)$$

Integrating (2.57) from a to t, we obtain

$$0 \le (\alpha(s))_a^t \le (-2\varepsilon)\int_a^t \left(\frac{-1}{\varphi(s)}\right)\exp\left\{-\int_a^s \frac{d\tau}{\varphi(\tau)}\right\}ds,$$

that is,

$$0 \le \alpha(t) - c \le (-2\varepsilon)\exp\left\{-\int_a^t \frac{d\tau}{\varphi(\tau)}\right\} + 2\varepsilon.$$

Hence,

$$0 \le (\alpha(t) - c)\exp\left\{\int_a^t \frac{d\tau}{\varphi(\tau)}\right\} \le 2\varepsilon \exp\left\{\int_a^t \frac{d\tau}{\varphi(\tau)}\right\} - 2\varepsilon$$

implies that

$$0 \le y(t) - \varepsilon - c\exp\left\{\int_a^t \frac{d\tau}{\varphi(\tau)}\right\} \le 2\varepsilon \exp\left\{\int_a^t \frac{d\tau}{\varphi(\tau)}\right\} - 2\varepsilon$$

due to (2.49), that is,

$$c\exp\left\{\int_a^t \frac{d\tau}{\varphi(\tau)}\right\} + \varepsilon \le y(t) \le 2\varepsilon \exp\left\{\int_a^t \frac{d\tau}{\varphi(\tau)}\right\} + c\exp\left\{\int_a^t \frac{d\tau}{\varphi(\tau)}\right\} - \varepsilon.$$

If $t \to a+$ in (2.49), then $c = y(a) - \varepsilon$. Therefore,

$$\begin{aligned}(y(a) - \varepsilon)\exp\left\{\int_a^t \frac{d\tau}{\varphi(\tau)}\right\} + \varepsilon \le y(t) &\le (2\varepsilon + c)\exp\left\{\int_a^t \frac{d\tau}{\varphi(\tau)}\right\} - \varepsilon \\ &\le (y(a) + \varepsilon)\exp\left\{\int_a^t \frac{d\tau}{\varphi(\tau)}\right\} - \varepsilon.\end{aligned}$$

Hence, the corollary is proved.

Corollary 17 *Assume that a is a real number and that $\varphi(t) < 0$ is true for any $t \in I$. Moreover, assume that $y : I \cup \{a\} \to \mathbb{R}$ is a function which is differentiable on I and continuous at a on the right. If y satisfies (2.48) for all $t \in I$ and for some $\varepsilon > 0$, then*

$$(y(a)+\varepsilon)\exp\left\{\int_a^t \frac{d\tau}{\varphi(\tau)}\right\}-\varepsilon \leq y(t) \leq (y(a)-\varepsilon)\exp\left\{\int_a^t \frac{d\tau}{\varphi(\tau)}\right\}+\varepsilon \quad (2.58)$$

for each $t \in I$.

Proof 50 *If y satisfies (2.48) for any $t \in I$, then Theorem 24 implies that there exists a differentiable function $\alpha : I \to \mathbb{R}$, satisfying (2.49) and (2.50) for all $t \in I$. Define $c = \lim_{t\to a+} \alpha(t)$, where $\alpha : I \to \mathbb{R}$ is given in Theorem 24. Since $\varphi(t) < 0$ for $t \in I$, then (2.50) becomes*

$$0 \geq \alpha'(t) \geq \frac{2\varepsilon}{\varphi(t)} \exp\left\{-\int_a^t \frac{d\tau}{\varphi(\tau)}\right\}. \quad (2.59)$$

Integrating (2.59) from a to t, we obtain

$$0 \geq (\alpha(s))_a^t \geq (-2\varepsilon)\int_a^t \left(\frac{-1}{\varphi(s)}\right) \exp\left\{-\int_a^s \frac{d\tau}{\varphi(\tau)}\right\} ds,$$

that is,

$$0 \geq \alpha(t) - c \geq 2\varepsilon - 2\varepsilon \exp\left\{-\int_a^t \frac{d\tau}{\varphi(\tau)}\right\}.$$

Hence,

$$0 \geq (\alpha(t)-c)\exp\left\{\int_a^t \frac{d\tau}{\varphi(\tau)}\right\} \geq 2\varepsilon \exp\left\{\int_a^t \frac{d\tau}{\varphi(\tau)}\right\} - 2\varepsilon$$

implies that

$$0 \geq y(t) - \varepsilon - c\exp\left\{\int_a^t \frac{d\tau}{\varphi(\tau)}\right\} \geq 2\varepsilon \exp\left\{\int_a^t \frac{d\tau}{\varphi(\tau)}\right\} - 2\varepsilon$$

due to (2.49), that is,

$$c\exp\left\{\int_a^t \frac{d\tau}{\varphi(\tau)}\right\} + \varepsilon \geq y(t) \geq (2\varepsilon + c)\exp\left\{\int_a^t \frac{d\tau}{\varphi(\tau)}\right\} - \varepsilon$$

implies that

$$(2\varepsilon + c)\exp\left\{\int_a^t \frac{d\tau}{\varphi(\tau)}\right\} - \varepsilon \leq y(t) \leq c\exp\left\{\int_a^t \frac{d\tau}{\varphi(\tau)}\right\} + \varepsilon.$$

If $t \to a+$ *in* (2.49), *then* $c = y(a) - \varepsilon$. *Therefore,*

$$(y(a) + \varepsilon) \exp\left\{ \int_a^t \frac{d\tau}{\varphi(\tau)} \right\} - \varepsilon \leq y(t) \leq (y(a) - \varepsilon) \exp\left\{ \int_a^t \frac{d\tau}{\varphi(\tau)} \right\} + \varepsilon.$$

Hence, the corollary is proved.

2.4 Stability of $p(x)y' - q(x)y - r(x) = 0$

Definition 10 *We say that* $p(x)y' - q(x)y - r(x) = 0$ *has the Hyers-Ulam stability if there exists a constant* $K \geq 0$ *with the following property: for every* $\epsilon > 0$, $y \in C^1(I)$, *if*

$$|p(x)y' - q(x)y - r(x)| \leq \epsilon,$$

then there exists some $z \in C^1(I)$ *satisfying* $p(x)z' - q(x)z - r(x) = 0$ *such that*

$$|y(x) - z(x)| \leq K\epsilon,$$

where $I = (a, b), -\infty \leq a \leq b \leq \infty$. *We call such* K *a Hyers-Ulam stability constant for the given equation.*

Theorem 26 *Let* $p(x), q(x)$ *and* $r(x)$ *be continuous real functions defined on the interval* $I = (a, b)$ *such that* $p(x) \neq 0$ *and* $|q(x)| \geq \delta$ *for all* $x \in I$ *and some* $\delta > 0$ *independent of* x. *Then* $p(x)y' - q(x)y - r(x) = 0$ *has the Hyers-Ulam stability.*

Proof 51 *Let* $\epsilon > 0$ *and* $y : I \to \mathbb{R}$ *be a continuously differentiable function such that*

$$|p(x)y' - q(x)y - r(x)| \leq \epsilon \tag{2.60}$$

holds for all $x \in I$. *We show that there exists a constant* $K > 0$ *independent of* ϵ, y *and* x *such that* $|y(x) - z(x)| \leq K\epsilon$ *for all* $x \in I$, *and for some* $z \in C^1(I)$ *satisfying* $p(x)z' - q(x)z - r(x) = 0, x \in I$. *Without loss of generality, we assume that* $q(x) \geq 1$ *for all* $x \in I$. *Let* $p(x) > 0$ *for all* $x \in I$. *In view of* (2.60), *we have*

$$-\epsilon \leq p(x)y' - q(x)y - r(x) \leq \epsilon$$

and hence

$$\begin{aligned} -\epsilon \frac{1}{p(x)} exp\Big\{ -\int_a^x \frac{q(s)}{p(s)} ds \Big\} &\leq \frac{1}{p(x)} exp\Big\{ -\int_a^x \frac{q(s)}{p(s)} ds \Big\} [p(x)y' - q(x)y - r(x)] \\ &\leq \epsilon \frac{1}{p(x)} exp\Big\{ -\int_a^x \frac{q(s)}{p(s)} ds \Big\}, \end{aligned}$$

that is,

$$-\epsilon\frac{q(x)}{p(x)}exp\Big\{-\int_a^x\frac{q(s)}{p(s)}ds\Big\} \le \frac{1}{p(x)}exp\Big\{-\int_a^x\frac{q(s)}{p(s)}ds\Big\}[p(x)y'-q(x)y-r(x)]$$
$$\le \epsilon\frac{q(x)}{p(x)}exp\Big\{-\int_a^x\frac{q(s)}{p(s)}ds\Big\},$$

that is,

$$-\epsilon\frac{q(x)}{p(x)}exp\Big\{-\int_a^x\frac{q(s)}{p(s)}ds\Big\} \le \frac{1}{p(x)}exp\Big\{-\int_a^x\frac{q(s)}{p(s)}ds\Big\}[p(x)y'-q(x)y]$$
$$-\frac{r(x)}{p(x)}exp\Big\{-\int_a^x\frac{q(s)}{p(s)}ds\Big\}$$
$$\le \epsilon\frac{q(x)}{p(x)}exp\Big\{-\int_a^x\frac{q(s)}{p(s)}ds\Big\}. \tag{2.61}$$

If we choose $b_1 \in [a,b]$ *such that* $y(b_1) < \infty$, *then for any* $x \in (a, b_1]$ *and integrating* (2.61) *from* x *to* b_1, *we get*

$$-\epsilon\left(exp\Big\{-\int_a^x\frac{q(s)}{p(s)}ds\Big\}-exp\Big\{-\int_a^{b_1}\frac{q(s)}{p(s)}ds\Big\}\right)$$
$$\le y(b_1)exp\Big\{-\int_a^{b_1}\frac{q(s)}{p(s)}ds\Big\}-y(x)exp\Big\{-\int_a^x\frac{q(s)}{p(s)}ds\Big\}$$
$$-\int_x^{b_1}\frac{r(s)}{p(s)}exp\Big\{-\int_a^s\frac{q(t)}{p(t)}dt\Big\}ds$$
$$\le \epsilon\left(exp\Big\{-\int_a^x\frac{q(s)}{p(s)}ds\Big\}-exp\Big\{-\int_a^{b_1}\frac{q(s)}{p(s)}ds\Big\}\right),$$

that is,

$$-\epsilon\, exp\Big\{-\int_a^x\frac{q(s)}{p(s)}ds\Big\}$$
$$\le (y(b_1)-\epsilon)exp\Big\{-\int_a^{b_1}\frac{q(s)}{p(s)}ds\Big\}-y(x)exp\Big\{-\int_a^x\frac{q(s)}{p(s)}ds\Big\}$$
$$-\int_x^{b_1}\frac{r(s)}{p(s)}exp\Big\{-\int_a^s\frac{q(t)}{p(t)}dt\Big\}ds$$
$$\le \epsilon\left(exp\Big\{-\int_a^x\frac{q(s)}{p(s)}ds\Big\}-2exp\Big\{-\int_a^{b_1}\frac{q(s)}{p(s)}ds\Big\}\right).$$

Therefore,

$$-\epsilon\, exp\Big\{-\int_a^x \frac{q(s)}{p(s)}ds\Big\}$$
$$\leq (y(b_1)-\epsilon)exp\Big\{-\int_a^{b_1}\frac{q(s)}{p(s)}ds\Big\} - y(x)exp\Big\{-\int_a^x \frac{q(s)}{p(s)}ds\Big\}$$
$$-\int_x^{b_1}\frac{r(s)}{p(s)}exp\Big\{-\int_a^s \frac{q(t)}{p(t)}dt\Big\}ds \leq \epsilon\, exp\Big\{-\int_a^x \frac{q(s)}{p(s)}ds\Big\}$$

implies that

$$-\epsilon \leq exp\Big\{\int_a^x \frac{q(s)}{p(s)}ds\Big\}\left[(y(b_1)-\epsilon)exp\Big\{-\int_a^{b_1}\frac{q(s)}{p(s)}ds\Big\}\right]$$
$$-exp\Big\{\int_a^x \frac{q(s)}{p(s)}ds\Big\}\left[\int_x^{b_1}\frac{r(s)}{p(s)}exp\Big\{-\int_a^s \frac{q(t)}{p(t)}dt\Big\}ds\right] - y(x) \leq \epsilon. \quad (2.62)$$

Similarly, for any $x \in [b_1, b)$ *and integrating* (2.61) *from* b_1 *to* x*, we obtain*

$$\epsilon \leq y(x) - exp\Big\{\int_a^x \frac{q(s)}{p(s)}ds\Big\}\left[(y(b_1)-\epsilon)exp\Big\{-\int_a^{b_1}\frac{q(s)}{p(s)}ds\Big\}\right]$$
$$+exp\Big\{\int_a^x \frac{q(s)}{p(s)}ds\Big\}\left[\int_x^{b_1}\frac{r(s)}{p(s)}exp\Big\{-\int_a^s \frac{q(t)}{p(t)}dt\Big\}ds\right]$$
$$\leq \epsilon\left(2exp\Big\{\int_{b_1}^x \frac{q(s)}{p(s)}ds\Big\}-1\right) \leq \epsilon\left(2exp\Big\{\int_a^b \frac{q(s)}{p(s)}ds\Big\}-1\right) = \epsilon(2A-1), \quad (2.63)$$

where $A = exp\Big\{\int_a^b \frac{q(s)}{p(s)}ds\Big\}$. *If we select*

$$z_1(x) = exp\Big\{\int_a^x \frac{q(s)}{p(s)}ds\Big\}\times$$
$$\times\left[(y(b_1)-\epsilon)exp\Big\{-\int_a^{b_1}\frac{q(s)}{p(s)}ds\Big\} - \int_x^{b_1}\frac{r(s)}{p(s)}exp\Big\{-\int_a^s \frac{q(t)}{p(t)}dt\Big\}ds\right],$$

then combining (2.62) *and* (2.63), *it follows that*

$$|y(x)-z_1(x)| \leq (2A-1)\epsilon,\ x \in I.$$

We note that

$$p(x)z_1' - q(x)z_1 - r(x) = 0\ x \in I.$$

For the case $p(x) < 0$, *we can apply the above similar argument and obtain that*

$$|y(x)-z_2(x)| \leq (2B-1)\epsilon,\ x \in I,$$

where $B = exp\Big\{ - \int_a^b \frac{q(s)}{p(s)} ds \Big\}$ *and*

$$z_2(x) = exp\Big\{ \int_a^x \frac{q(s)}{p(s)} ds \Big\} \times$$

$$\times \left[(y(b_1) - \epsilon) exp\Big\{ - \int_a^{b_1} \frac{q(s)}{p(s)} ds \Big\} - \int_x^{b_1} \frac{r(s)}{p(s)} exp\Big\{ - \int_a^s \frac{q(t)}{p(t)} dt \Big\} ds \right].$$

Indeed,

$$p(x)z_2' - q(x)z_2 - r(x) = 0 \ x \in I.$$

2.5 Stability of $y' = \lambda y$ on Banach Spaces

In this section, we consider the Hyers-Ulam-Rassias stability of the Banach space valued differential equation $y' = \lambda y$, where λ is a complex constant.

Let $(X, ||.||)$ be a non-zero complex Banach space and $I = (a, b)$ be an open interval in extended real line system. A function $f(t) \in X$ is said to be strongly differentiable, if for all $t \in I$ there exists a function $f'(t) \in X, \ \forall t \in I$ such that

$$\lim_{s \to 0} \left|\left| \frac{f(t+s) - f(t)}{s} - f'(t) \right|\right| = 0.$$

Consider the following two equivalent statements:
(i) $f'(t) = \lambda f(t), \ \forall t \in I.$
(ii) There is an $x \in X$ such that $f(t) = e^{\lambda t} x, \ \forall t \in I.$

Theorem 27 *Suppose* λ *is a complex number,* $\varepsilon : I \to [0, \infty)$ *is a continuous function and* $f : I \to X$ *is a strongly differentiable function such that*

$$||f'(t) - \lambda f(t)|| \leq \epsilon(t) \tag{2.64}$$

for all $t \in I$.

(a) *If* $\epsilon(t)e^{-Re(\lambda)t}$ *is integrable on* $(a, t_a]$ *for some* $t_a \in I$*, then there is a unique* $x_a \in X$ *such that*

$$||f(t) - e^{\lambda t} x_a|| \leq e^{Re(\lambda)t} \int_a^t \epsilon(\sigma) e^{-Re(\lambda)\sigma} d\sigma$$

for all $t \in I$.

(b) *If* $\epsilon(t)e^{-Re(\lambda)t}$ *is integrable on* $[t_b, b)$ *for some* $t_b \in I$*, then there is a unique* $x_b \in X$ *such that*

$$||f(t) - e^{\lambda t} x_b|| \leq e^{Re(\lambda)t} \int_t^b \epsilon(\sigma) e^{-Re(\lambda)\sigma} d\sigma$$

for all $t \in I$.

Proof 52 *Let X^* be the dual space of X and $f_\varphi : I \to \mathbb{C}$, defined by $f_\varphi(t) = \varphi(f(t)), \forall t \in I$, $\varphi \in X^*$ and $||x|| = \sup\{|\psi(x)| : \psi \in X^*, ||\psi|| = 1, \forall x \in X\}$. $(f_\varphi)'(t) = \varphi f'(t)$ and hence φ is continuous. Now,*

$$\begin{aligned}|(f_\varphi)'(t) - \lambda f_\varphi(t)| &= |\varphi f'(t) - \varphi(\lambda f(t))| \\ &\leq ||\varphi||\, ||((f)'(t)) - (\lambda f(t))|| \\ &\leq ||\varphi|| \epsilon(t)\end{aligned}$$

implies that

$$\begin{aligned}\left| \int_s^t \{e^{-\lambda\sigma} f_\varphi(\sigma)\}' d\sigma \right| &= \left| \int_s^t \{(f_\varphi)'(\sigma)e^{-\lambda\sigma} - \lambda f_\varphi(\sigma)e^{-\lambda\sigma}\} d\sigma \right| \\ &= \left| \int_s^t \{(f_\varphi)'(\sigma) - \lambda f_\varphi(\sigma)\} e^{-\lambda\sigma} d\sigma \right| \\ &\leq ||\varphi|| \left| \int_s^t \epsilon(\sigma) e^{-Re(\lambda)\sigma} d\sigma \right|, \forall s, t \in I. \qquad (2.65)\end{aligned}$$

Hence,

$$|\varphi(e^{-\lambda t} f(t) - e^{-\lambda s} f(s))| = |e^{-\lambda t} f_\varphi(t) - e^{-\lambda s} f_\varphi(s))| \leq ||\varphi|| \left| \int_s^t \epsilon(\sigma) e^{-Re(\lambda)\sigma} d\sigma \right| \qquad (2.66)$$

for all $s, t \in I$. Therefore,

$$\begin{aligned}||e^{-\lambda t} f(t) - e^{-\lambda s} f(s)|| &= \sup\{|\psi(e^{-\lambda t} f(t) - e^{-\lambda s} f(s))| : \psi \in X^*, ||\psi|| = 1\} \\ &\leq \sup\left\{ ||\psi|| \left| \int_s^t \epsilon(\sigma) e^{-Re(\lambda)\sigma} d\sigma \right| : \psi \in X^*, ||\psi|| = 1 \right\} \\ &= \left| \int_s^t \epsilon(\sigma) e^{-Re(\lambda)\sigma} d\sigma \right|, \forall s, t \in I. \qquad (2.67)\end{aligned}$$

(a) By (2.67) it follows that $\{e^{-\lambda s} f(s)\}_{s \in I}$ is a Cauchy net and so $\{e^{-\lambda s} f(s)\}$ converges to an element $x_a \in X$ when $s \to a+$. Now,

$$\begin{aligned}||f(t) - e^{\lambda t} x_a|| &= ||f(t) - e^{\lambda(t-s)} f(s) + e^{\lambda(t-s)} f(s) - e^{\lambda t} x_a|| \\ &\leq ||f(t) - e^{\lambda(t-s)} f(s)|| + ||e^{\lambda(t-s)} f(s) - e^{\lambda t} x_a|| \\ &= ||e^{\lambda t}(e^{-\lambda t} f(t) - e^{-\lambda s} f(s))|| + ||e^{\lambda t}(e^{-\lambda s} f(s) - x_a)|| \\ &= |e^{\lambda t}|\, ||(e^{-\lambda t} f(t) - e^{-\lambda s} f(s))|| + |e^{\lambda t}|\, ||(e^{-\lambda s} f(s) - x_a)|| \\ &\leq e^{Re(\lambda)t} \left| \int_s^t \epsilon(\sigma) e^{-Re(\lambda)\sigma} d\sigma \right| + e^{Re(\lambda)t} ||e^{-\lambda s} f(s) - x_a||, \forall s, t \in I \qquad (2.68)\end{aligned}$$

due to (2.67). As $s \to a+$, $e^{-\lambda s} f(s) \to x_a$. Hence,

$$||f(t) - e^{\lambda t} x_a|| \leq e^{Re(\lambda)t} \int_a^t \epsilon(\sigma) e^{-Re(\lambda)\sigma} d\sigma, \forall t \in I. \qquad (2.69)$$

For any $x \in X$, let's suppose that

$$||f(t) - e^{\lambda t}x|| \leq e^{Re(\lambda)t} \int_a^t \epsilon(\sigma)e^{-Re(\lambda)\sigma}d\sigma, \forall t \in I.$$

Then

$$\begin{aligned}
||x_a - x|| &= ||x_a - e^{-\lambda t}f(t) + e^{-\lambda t}f(t) - x|| \\
&\leq ||x_a - e^{-\lambda t}f(t)|| + ||e^{-\lambda t}f(t) - x|| \\
&= ||e^{-\lambda t}(e^{\lambda t}x_a - f(t))|| + ||e^{-\lambda t}(f(t) - e^{\lambda t}x)|| \\
&= |e^{-\lambda t}|\,||e^{\lambda t}x_a - f(t)|| + |e^{-\lambda t}|\,||f(t) - e^{\lambda t}x|| \\
&\leq e^{-Re(\lambda)t}e^{Re(\lambda)t}\int_a^t \epsilon(\sigma)e^{-Re(\lambda)\sigma}d\sigma + e^{-Re(\lambda)t}e^{Re(\lambda)t} \\
&\quad \times \int_a^t \epsilon(\sigma)e^{-Re(\lambda)\sigma}d\sigma \\
&= 2\int_a^t \epsilon(\sigma)e^{-Re(\lambda)\sigma}d\sigma \\
&\to 0 \text{ when } t \to a+.
\end{aligned}$$

Consequently, x_a is unique.

(*b*) *Now suppose that $\epsilon(t)e^{-Re(\lambda)t}$ is integrable on $[t_b, b)$ for some $t_b \in I$. By* (*a*), *$\epsilon(t)e^{-Re(\lambda)t}$ is integrable on $[t_0, b), \forall t_0 \in I$. By* (2.67), *$e^{-\lambda s}f(s)$ converges to x_b such that $x_b \in X$ when $s \to b-$. By* (2.65), *we obtain*

$$\left|\int_t^s \{e^{-\lambda\sigma}f_\varphi(\sigma)\}'d\sigma\right| \leq ||\varphi||\left|\int_t^s \epsilon(\sigma)e^{-Re(\lambda)(\sigma)}d\sigma\right|, \forall s, t \in I. \tag{2.70}$$

Since

$$\int_t^s \{e^{-\lambda\sigma}f_\varphi(\sigma)\}'d\sigma = \varphi(e^{-\lambda s}f(s) - e^{-\lambda t}f(t)) \tag{2.71}$$

for all $s, t \in I$, then from (2.70) *and* (2.71) *we find*

$$||\varphi(e^{-\lambda s}f(s) - e^{-\lambda t}f(t))|| \leq ||\varphi||\left|\int_t^s \epsilon(\sigma)e^{-Re(\lambda)\sigma}d\sigma\right| \tag{2.72}$$

for all $s, t \in I$. As a result,

$$||e^{-\lambda t}f(t) - e^{-\lambda s}f(s)|| \leq \left|\int_t^s \epsilon(\sigma)e^{-Re(\lambda)\sigma}d\sigma\right|$$

for all $s, t \in I$. Taking $s \to b-$ and proceeding as in (2.68), *it follows that*

$$||f(t) - e^{\lambda t}x_b|| \leq e^{Re(\lambda)t}\left|\int_t^s \epsilon(\sigma)e^{-Re(\lambda)\sigma}d\sigma\right| + e^{Re(\lambda)t}||e^{-\lambda s}f(s) - x_b||, \forall s, t \in I.$$

Since $e^{-\lambda s}f(s) \to x_b$ *as* $s \to b+$, *then*

$$||f(t) - e^{\lambda t}x_b|| \leq e^{Re(\lambda)t}\left|\int_t^b \epsilon(\sigma)e^{-Re(\lambda)\sigma}d\sigma\right|, \forall t \in I.$$

Using same type of reasoning as above, we can prove that x_b *is unique. Hence, the theorem is proved.*

Corollary 18 *Let* $f : I \to X$ *be a strongly differentiable function that satisfies the inequality*

$$||f'(t) - \lambda f(t)|| \leq \varepsilon \tag{2.73}$$

for all $t \in I$ *and for some* $\varepsilon > 0$. *If* $Re(\lambda) \neq 0$ *and* $m = 0$, *there exists a unique* $x_0 \in X$ *such that*

$$\sup_{t\in I} ||f(t) - e^{\lambda t}x_0|| < \infty.$$

Moreover, for the above $x_0 \in X$, *the following estimate*

$$||f(t) - e^{\lambda t}x_0|| \leq \frac{\varepsilon}{|Re(\lambda)|} \tag{2.74}$$

holds for all $t \in I$, *where* $m = \inf\{e^{-Re(\lambda)t} : t \in I\} = \lim_{\sigma\to a+} e^{-Re(\lambda)\sigma}$.

Proof 53 *For* $Re(\lambda) < 0, \varepsilon e^{-Re(\lambda)t}$ *is integrable on* $(a, t_a]$ *for any* $t_a \in I$. *By Theorem 27(a), there is an* $x_a \in X$ *such that*

$$||f(t) - e^{\lambda t}x_a|| \leq \varepsilon e^{Re(\lambda)t}\int_a^t e^{-Re(\lambda)\sigma}d\sigma = \frac{\varepsilon}{|Re(\lambda)|} - m = \frac{\varepsilon}{|Re(\lambda)|}$$

for all $t \in I$.

Again, for $Re(\lambda) > 0, \varepsilon e^{-Re(\lambda)t}$ *is integrable on* $[t_b, b)$ *for any* $t_b \in I$. *By Theorem 27(b), there exists an* $x_b \in X$ *such that*

$$||f(t) - e^{\lambda t}x_b|| \leq \varepsilon e^{Re(\lambda)t}\int_t^b e^{-Re(\lambda)\sigma}d\sigma = m - \frac{\varepsilon}{Re(\lambda)} = \frac{\varepsilon}{|Re(\lambda)|}$$

for all $t \in I$. *Define*

$$x_0 = \begin{cases} x_a, & for\ Re(\lambda) < 0, \\ x_b, & for\ Re(\lambda) > 0. \end{cases}$$

Clearly, x_0 *satisfies* (2.74) *for all* $t \in I$. *Let* $x_1 \in X$ *such that* $||f(t) - e^{\lambda t}x_1|| \leq L$, $\forall t \in I$ *for some* $0 < L < \infty$. *By* (2.74),

$$\begin{aligned}
||x_0 - x_1|| &= ||x_0 - e^{-\lambda t}f(t) + e^{-\lambda t}f(t) - x_1|| \\
&\leq ||x_0 - e^{-\lambda t}f(t)|| + ||e^{-\lambda t}f(t) - x_1|| \\
&= ||e^{-\lambda t}||\ ||e^{\lambda t}x_0 - f(t)|| + ||e^{-\lambda t}||\ ||f(t) - e^{\lambda t}x_1|| \\
&\leq e^{-Re(\lambda)t}\left(\frac{\varepsilon}{|Re(\lambda)|} + L\right)
\end{aligned}$$

for all $t \in I$. *As* $m = \inf\{e^{-Re(\lambda)t} : t \in I\} = 0$, x_0 *is unique.*

Corollary 19 *Assume that a strongly differentiable function $f : I \to X$ satisfies (2.73) for all $t \in I$ and for some $\varepsilon > 0$. If $Re(\lambda) \neq 0$ and $m > 0$, then there are infinitely many $x_0 \in X$ for which the inequality*

$$||f(t) - e^{\lambda t}x_0|| \leq \frac{\varepsilon}{|Re(\lambda)|}\left(1 - \frac{m}{M}\right) \tag{2.75}$$

holds for all $t \in I$. More explicitly, if S is the set of all $x_0 \in X$ satisfying (2.75), then the cardinal number of S is at least c, where c denotes that of the continuum and $M = \sup\{e^{-Re(\lambda)t} : t \in I\}$.

Proof 54 *As in Corollary 18 and Theorem 27, if we define*

$$x_0 = \begin{cases} \lim_{s\to a+} e^{-\lambda s} f(s), & for \ Re(\lambda) < 0, \\ \lim_{s\to b-} e^{-\lambda s} f(s), & for \ Re(\lambda) > 0, \end{cases}$$

then (2.75) is true for all $t \in I$. Since

$$m = \begin{cases} \lim_{s\to a+} e^{-\lambda s} s, & for \ Re(\lambda) < 0, \\ \lim_{s\to b-} e^{-\lambda s} s, & for \ Re(\lambda) > 0, \end{cases}$$

then we define

$$J = \{e^{-Re(\lambda)t} : m < e^{-Re(\lambda)s} \leq e^{-Re(\lambda)t} < M \ implies \ e^{-\lambda s} f(s) = x_0\}.$$

We may note that either $J = \phi$ or $J \neq \phi$ with $supJ < M$. Define

$$\alpha = \begin{cases} supJ, & for \ J \neq \phi, \\ m, & for \ J = \phi. \end{cases}$$

By the definitions of J and α, $e^{-\lambda s} f(s) \to x_0$ when $e^{-Re(\lambda)s} \to \alpha$, and therefore there is $0 < \delta_0 < \alpha\left(1 - \frac{m}{M}\right)$ such that $\alpha < e^{-Re(\lambda)s} < \alpha + \delta_0$ implies that

$$|e^{-Re(\lambda)s} - \alpha| \leq |\delta_0|.$$

Hence,

$$\begin{aligned} ||x_0 - e^{-\lambda s} f(s)|| &\leq \left|\alpha\left(1 - \frac{m}{M}\right)\right| \\ &\leq \frac{m\varepsilon}{|Re(\lambda)|}\left(1 - \frac{m}{M}\right), \end{aligned} \tag{2.76}$$

where $\alpha = \frac{m\varepsilon}{|Re(\lambda)|}$. By the definition of J, there exists an $s_0 \in I$ such that

$$\alpha < e^{-Re(\lambda)s_0} < \alpha + \delta_0 \tag{2.77}$$

and $x_0 \neq e^{-\lambda s_0} f(s_0)$. Consider $r_0 = ||x_0 - e^{-\lambda s_0} f(s_0)|| > 0$. The function

$s \mapsto ||x_0 - e^{-\lambda s} f(s)||$ is continuous so, by the intermediate value theorem, there exists to each $r \in (0, r_0)$ an $s_r \in I$ such that

$$\alpha < e^{-Re(\lambda)s_r} < e^{-Re(\lambda)s_0} \tag{2.78}$$

and

$$||x_0 - e^{-\lambda s_r} f(s_r)|| = r. \tag{2.79}$$

Let $x_r = e^{-\lambda s_r} f(s_r)$, $\forall r \in (0, r_0)$. If $r_1, r_2 \in (0, r_0)$ and $r_1 \neq r_2$, then $x_{r_1} \neq x_{r_2}$ by (2.79). We obtain from (2.76), (2.77) and (2.78) that

$$||x_0 - e^{-\lambda s_r} f(s_r)|| \leq \frac{m\varepsilon}{|Re(\lambda)|}\left(1 - \frac{m}{M}\right), \ \forall r \in (0, r_0). \tag{2.80}$$

Choose $t \in I$ arbitrarily. Now, we have two possibilities: either $e^{-Re(\lambda)t} \leq \alpha$ or $e^{-Re(\lambda)t} > \alpha$. In the former case and by the definition of J, we obtain $e^{-\lambda t} f(t) = x_0$. Therefore, (2.80) becomes

$$\begin{aligned} ||f(t) - e^{\lambda t} x_r|| &= e^{Re(\lambda)t} ||x_0 - e^{-\lambda s_r} f(s_r)|| \\ &\leq \frac{\varepsilon}{|Re(\lambda)|}\left(1 - \frac{m}{M}\right), \ \forall\, r \in (0, r_0). \end{aligned}$$

In the latter case as $0 < \delta_0 < \alpha\left(1 - \frac{m}{M}\right)$, (2.77) and (2.78) give rise to

$$\alpha < e^{-Re(\lambda)s_r} < \alpha\left(2 - \frac{m}{M}\right), \ \forall r \in (0, r_0). \tag{2.81}$$

Ultimately, (2.81) becomes

$$\frac{\alpha}{e^{-Re(\lambda)t}} < \frac{e^{-Re(\lambda)s_r}}{e^{-Re(\lambda)t}} < \frac{\alpha}{e^{-Re(\lambda)t}}\left(2 - \frac{m}{M}\right),$$

that is,

$$\frac{m}{M} < \frac{e^{-Re(\lambda)s_r}}{e^{-Re(\lambda)t}} < 2 - \frac{m}{M}$$

due to $m \leq \alpha < e^{-Re(\lambda)t} < M$. Therefore,

$$\left|\frac{e^{-Re(\lambda)s_r}}{e^{-Re(\lambda)t}} - 1\right| \leq 1 - \frac{m}{M}, \ \forall\, r \in (0, r_0). \tag{2.82}$$

By (2.67),

$$||e^{-\lambda t} f(t) - e^{-\lambda s_r}|| \leq \frac{\varepsilon}{|Re(\lambda)|} |e^{-Re(\lambda)t} - e^{-Re(\lambda)s_r}|, \ \forall\, r \in (0, r_0). \tag{2.83}$$

From (2.82) and (2.83), it follows that

$$\begin{aligned} ||f(t) - e^{\lambda t} x_r|| &\leq e^{Re(\lambda)t} \frac{\varepsilon}{|Re(\lambda)|} |e^{-Re(\lambda)t} - e^{-Re(\lambda)s_r}| \\ &\leq \frac{\varepsilon}{|Re(\lambda)|}\left(1 - \frac{m}{M}\right), \ \forall\, r \in (0, r_0). \end{aligned}$$

Hence, all elements of $\{x_r : r \in (0, r_0)\}$ satisfy (2.75) for each $t \in I$. Therefore, by (2.79), the cardinal number of the set $\{x_r : r \in (0, r_0)\}$ is that of $(0, r_0)$, and hence c. Thus, the corollary is proved.

Corollary 20 *Assume that a strongly differentiable function $f : I \to X$ satisfies the inequality (2.73), for all $t \in I$ and for some $\varepsilon > 0$. If $Re(\lambda) = 0$ and the diameter $\delta(I)$ of I is finite, then there exist unique $x_a \in X$ and $x_b \in X$ such that*

$$||f(t) - e^{\lambda t}x_a|| \leq \varepsilon(t-a) \text{ and } ||f(t) - e^{\lambda t}x_b|| \leq \varepsilon(b-t)$$

for all $t \in I$ respectively.

Proof 55 *As $\delta(I)$ is finite, ε is integrable on (a, b). Here $Re(\lambda) = 0$. Hence, by (a) and (b) of Theorem 27, there exist unique $x_a \in X$ and $x_b \in X$ such that*

$$||f(t) - e^{\lambda t}x_a|| \leq \varepsilon(t-a) \text{ and } ||f(t) - e^{\lambda t}x_b|| \leq \varepsilon(b-t)$$

for all $t \in I$ respectively. Hence, the corollary is proved.

2.6 Stability of $y' = F(x, y)$

In this section, for a bounded and continuous function $F(x, y)$, we will prove the Hyers-Ulam-Rassias stability as well as the Hyers-Ulam stability of the differential equations of the form

$$y'(x) = F(x, y(x)). \tag{2.84}$$

Theorem 28 *Let (X, d) be a generalized complete metric space. Assume that $\Lambda : X \to X$ is a strictly contractive operator with the Lipschitz constant $L < 1$. If there exists a nonnegative integer k such that $d(\Lambda^{k+1}x, \Lambda^k x) < \infty$ for some $x \in X$, then the followings are true:*

(a) *The sequence $\Lambda^n x$ converges to a fixed point x^* of Λ;*

(b) *x^* is the unique fixed point of Λ in*

$$X^* = \{y \in X \mid d(\Lambda^k x, y) < \infty\};$$

(c) *If $y \in X^*$, then*

$$d(y, x^*) \leq \frac{1}{1-L} d(\Lambda, y).$$

Proof 56 *Let $x_0 \in X$, and consider that: sequence of successive approximations with initial element $x_0 : x_0, \Lambda x_0, \Lambda^2 x_0, \cdots, \Lambda^l x_0, \cdots$. Then the following alternative holds such that either*
(A) for each integer $l = 0, 1, 2, \cdots$, we have

$$d(\Lambda^l x_0, \Lambda^{l+1}x_0) = \infty,$$

or
(B) the sequence of successive approximations $x_0, \Lambda x_0, \Lambda^2 x_0, \cdots, \Lambda^l x_0, \cdots$, *is* $d-$*convergent to a fixed point of* Λ.
(a) Assume that sequence of numbers $d(x_0, \Lambda x_0), d(\Lambda x_0, \Lambda^2 x_0), \cdots$, $d(\Lambda^l x_0, \Lambda^{l+1} x_0), \cdots$, *the sequence of distances between consecutive neighbors of the sequence of successive approximations with initial element* x_0. *There are two mutually exclusive possibilities: either*
(i) for each integer $l = 0, 1, 2, \cdots$, *we have either*

$$d(\Lambda^l x_0, \Lambda^{l+1} x_0) = \infty,$$

or
(ii) for some integer $l = 0, 1, 2, \cdots$, *one has*

$$d(\Lambda^l x_0, \Lambda^{l+1} x_0) < \infty.$$

In case(ii), let $\mathbb{N} = \mathbb{N}(x_0)$ *denote a particular one of all the integers* $l = 0, 1, 2, \cdots$ *such that*

$$d(\Lambda^l x_0, \Lambda^{l+1} x_0) < \infty.$$

Since $d(\Lambda^{\mathbb{N}} x_0, \Lambda^{\mathbb{N}+1} x_0) < \infty$ *and* Λ *is a strictly contractive operator with the Lipschitz constant, so by the given theorem we obtain that*

$$d(\Lambda^{n+l} x_0, \Lambda^{\mathbb{N}+l+1} x_0) = d(\Lambda\Lambda^{n+l-1} x_0, \Lambda\Lambda^{\mathbb{N}+l} x_0)$$

Theorem 29 *For given real numbers* a *and* b *with* $a < b$, *let* $I = [a, b]$ *be a closed interval and choose a* $c \in I$. *Let* K *and* L *be positive constants with* $0 < KL < 1$. *Assume that* $F : I \times \mathbb{R} \to \mathbb{R}$ *is a continuous function which satisfies a Lipschitz condition*

$$|F(x, y) - F(x, z)| \leq L|y - z| \tag{2.85}$$

for any $x \in I$ *and* $y, z \in \mathbb{R}$. *If a continuously differentiable function* $y : I \to \mathbb{R}$ *satisfies*

$$|y'(x) - F(x, y(x))| \leq \varphi(x) \tag{2.86}$$

for all $x \in I$, *where* $\varphi : I \to (0, \infty)$ *is a continuous function with*

$$\left| \int_c^x \varphi(\tau) d\tau \right| \leq K\varphi(x) \tag{2.87}$$

for each $x \in I$, *then there exists a unique continuous function* $y_0 : I \to \mathbb{R}$ *such that*

$$y_0(x) = y(c) + \int_c^x F(\tau, y_0(\tau)) d\tau \tag{2.88}$$

(consequently, y_0 *is a solution to (2.84)) and*

$$|y(x) - y_0(x)| \leq \frac{K}{1 - KL}\varphi(x) \tag{2.89}$$

for all $x \in I$.

Proof 57 *Define a set X of all continuous functions $f : I \to \mathbb{R}$ by*

$$X = \{f : I \to \mathbb{R} : f \ \text{ is continuous}\} \tag{2.90}$$

and also introduce a generalized metric on X as given below:

$$d(f, g) = \inf\{C \in [0, \infty] : |f(x) - g(x)| \le C\varphi(x), \forall x \in I\}. \tag{2.91}$$

Assume that $d(f, g) > d(f, h) + d(h, g)$ is true for some $f, g, h \in X$. Hence, by (2.91), there exists an $x_0 \in I$ such that

$$\begin{aligned} |f(x_0) - g(x_0)| &> \{d(f, h) + d(h, g)\}\varphi(x_0) \\ &= d(f, h)\varphi(x_0) + d(h, g)\varphi(x_0) \\ &\ge |f(x_0) - h(x_0)| + |h(x_0) - g(x_0)|, \end{aligned}$$

which is a contradiction to the triangle law.

Let's assume that (X, d) be complete. Suppose $\{h_n\}$ be a Cauchy sequence in (X, d). So, for any $\varepsilon > 0$, there exists an integer $\mathbb{N}_\epsilon > 0$ such that $d(h_m, h_n) \le \varepsilon$ for each $m, n \ge \mathbb{N}_\varepsilon$. By (2.91) we obtain that

$$|h_m(x) - h_n(x)| \le \varepsilon\varphi(x) \tag{2.92}$$

for all $\varepsilon > 0$ and there exists $\mathbb{N}_\varepsilon \in \mathbb{N}$ for all $m, n \ge \mathbb{N}_\varepsilon \forall x \in I$. By fixing x, (2.92) implies that $\{h_n(x)\}$ is a Cauchy sequence in $\mathbb{R}$. Since $\mathbb{R}$ is complete, then $h_n(x)$ converges for all $x \in I$. Hence, we can define a function $h : I \to \mathbb{R}$ by

$$h(x) = \lim_{n\to\infty} h_n(x).$$

For $m \to \infty$, (2.92) implies that

$$\forall\, \varepsilon > 0\, \exists\, \mathbb{N}_\varepsilon \in \mathbb{N}\, \forall n \ge \mathbb{N}_\varepsilon\, \forall x \in I : |h(x) - h_n(x)| \le \varepsilon\varphi(x). \tag{2.93}$$

But φ is bounded on I, so $\{h_n\}$ converges uniformly to h. Therefore, h is continuous and $h \in X$.

Again (2.91) and (2.93) imply that

$$\forall\, \varepsilon > 0\, \exists\, \mathbb{N}_\varepsilon \in \mathbb{N}\, \forall n \ge \mathbb{N}_\varepsilon\, \forall x \in I : d(h, h_n) \le \varepsilon.$$

So, the Cauchy sequence $\{h_n\}$ converges to h in (X, d). Hence, (X, d) is complete. Consider, an operator $\Lambda : X \to X$ defined by

$$(\Lambda f)(x) = y(c) + \int_c^x F(\tau, f(\tau))d\tau,\ x \in I,\ \forall f \in X. \tag{2.94}$$

For each $f, g \in X$, suppose $C_{fg} \in [0, \infty]$ be an arbitrary constant such that $d(f, g) \le C_{fg}$. Hence, (2.91) implies that

$$|f(x) - g(x)| \le C_{fg}\varphi(x)\ \forall x \in I. \tag{2.95}$$

By (2.85), (2.87), (2.91), (2.94) *and* (2.95) *we obtain that*

$$\begin{aligned}|(\Lambda f)(x) - (\Lambda g)(x)| &= \left|\int_c^x \{F(\tau, f(\tau)) - F(\tau, g(\tau))\}d\tau\right| \\ &\leq \left|\int_c^x |F(\tau, f(\tau)) - F(\tau, g(\tau))|d\tau\right| \\ &\leq L\left|\int_c^x |f(\tau) - g(\tau)|d\tau\right| \\ &\leq LC_{fg}\left|\int_c^x \varphi(\tau)d\tau\right| \\ &\leq KLC_{fg}\varphi(x), \ \forall x \in I\end{aligned}$$

implies that $d(\Lambda f, \Lambda g) \leq KLC_{fg}$. *Ultimately,* $d(\Lambda f, \Lambda g) \leq KLd(f, g)$, $\forall f, g \in X$, *where* $0 < KL < 1$.

For an arbitrary $g_0 \in X$, *there exists a constant* $0 < C < \infty$ *such that it can be concluded from* (2.90) *and* (2.94) *that*

$$|(\Lambda g_0)(x) - g_0(x)| = \left|y(c) + \int_c^x F(\tau, g_0(\tau))d\tau - g_0(x)\right|, \ \forall x \in I. \quad (2.96)$$

But by (2.95), $|(\Lambda g_0)(x) - g_0(x)| \leq C\varphi(x)$. *Hence,* $\left|y(c) + \int_c^x F(\tau, g_0(\tau))d\tau - g_0(x)\right| \leq C\varphi(x) < \infty \ \forall x \in I$ *by* (2.96), *that is,* $d(\Lambda g_0, g_0) < \infty$.

Hence, by Theorem 28 (a), there exists a continuous function $y_0 : I \to \mathbb{R}$ *such that* $\Lambda^n g_0 \to y_0$ *in* (X, d) *and* $\Lambda y_0 = y_0$, *and hence,* y_0 *satisfies equation* (2.88) *for all* $x \in I$.

Let $g \in X$ *be arbitrary. As* g *and* g_0 *are bounded on* I *and* $\min_{x \in I} \varphi(x) > 0$, *so there exists a constant* $0 < C_g < \infty$ *such that*

$$|g_0(x) - g(x)| \leq C_g\varphi(x), \ \forall x \in I.$$

Hence, $d(g_0, g) < \infty$, *for each* $g \in X$ *and* $\{g \in X | d(g_0, g) < \infty\} = X$. *By Theorem 28(b), we conclude that* y_0 *is the unique continuous function by using the property of* (2.88). (2.88) *implies that*

$$-\varphi(x) \leq y'(x) - F(x, y(x)) \leq \varphi(x), \ \forall x \in I. \quad (2.97)$$

Integrating (2.97) *from* c *to* x, *we get*

$$\left|y(x) - y(c) - \int_c^x F(\tau, y(\tau))d\tau\right| \leq \left|\int_c^x \varphi(\tau))d\tau\right|, \ \forall x \in I.$$

From (2.87) *and* (2.94), *we obtain*

$$|y(x) - (\Lambda y)(x)| \leq \left|\int_c^x \varphi(\tau)d\tau\right| \leq K\varphi(x), \ \forall x \in I$$

implies that

$$d(y, \Lambda y) \leq K. \tag{2.98}$$

Ultimately, from Theorem 28(c) and (2.98) we obtain

$$d(y, y_0) \leq \frac{d(\Lambda y, y)}{1 - KL} \leq \frac{K}{1 - KL},$$

which is (2.89) for all $x \in I$.

Theorem 30 *For given real numbers a and b, let I denote either $(-\infty, b]$ or $\mathbb{R}$ or $[a, \infty)$. Set either $c = a$ for $I = [a, \infty)$ or $c = b$ for $I = (-\infty, b]$ or c is a fixed real number if $I = \mathbb{R}$. Let K and L be positive constants with $0 < KL < 1$. Assume that $F : I \times \mathbb{R} \to \mathbb{R}$ is a continuous function which satisfies a Lipschitz condition (2.85) for all $x \in I$ and all $y, z \in \mathbb{R}$. If a continuously differentiable function $y : I \to \mathbb{R}$ satisfies the differential inequality (2.86) for all $x \in I$, where $\varphi : I \to (0, \infty)$ is a continuous function satisfying the condition (2.87) for any $x \in I$, then there exists a unique continuous function $y_0 : I \to \mathbb{R}$ which satisfies (2.88) and (2.89) for all $x \in I$.*

Proof 58 *Define $I_n = [c - n, c + n]$, $\forall n \in \mathbb{N}$. By Theorem 29, there exists a unique continuous function $y_n : I_n \to \mathbb{R}$ such that*

$$y_n(x) = y(c) + \int_c^x F(\tau, y_n(\tau)) d\tau \tag{2.99}$$

and

$$|y(x) - y_n(x)| \leq \frac{K}{1 - KL} \varphi(x), \ \forall x \in I_n. \tag{2.100}$$

By the uniqueness of y_n if $x \in I_n$, then

$$y_n(x) = y_{n+1}(x) = y_{n+2}(x) = \dots . \tag{2.101}$$

For each $x \in \mathbb{R}$, define $n(x) \in \mathbb{N}$ by

$$n(x) = \min\{n \in \mathbb{N} : x \in I_n\}.$$

Define a function $y_0 : \mathbb{R} \to \mathbb{R}$ as

$$y_0(x) = y_{n(x)}(x). \tag{2.102}$$

We claim that y_0 is continuous. Let $x_1 \in \mathbb{R}$ be arbitrary and choose the integer $n_1 = n(x_1)$. Hence, x_1 belongs to the interior of I_{n_1+1} and there exists an $\varepsilon > 0$ such that $y_0(x) = y_{n_1+1}(x)$ for each x with $x_1 - \varepsilon < x < x_1 + \varepsilon$. But y_{n_1+1} is continuous at x_1, so is y_0. Therefore, y_0 is continuous at x_1 for all $x_1 \in \mathbb{R}$.

Let $x \to \mathbb{R}$ be arbitrary and choose the integer $n(x)$. Then, for $x \in I_{n(x)}$, it follows from (2.99) and (2.102) that

$$y_0(x) = y_{n(x)}(x) = y(c) + \int_c^x F(\tau, y_{n(x)}(\tau)) d\tau = y(c) + \int_c^x F(\tau, y_0(\tau)) d\tau.$$

By (2.101) and (2.102) we obtain that

$$y_{n(x)}(\tau) = y_{n(\tau)}(\tau) = y_0(\tau).$$

By (2.100) and (2.102), we get

$$|y(x) - y_0(x)| = |y(x) - y_{n(x)}(x)| \leq \frac{K}{1-KL}\varphi(x),$$

where $x \in I_{n(x)}$ *for any* $x \in \mathbb{R}$.

For uniqueness of y_0 *assume that* $z_0 : \mathbb{R} \to \mathbb{R}$ *is another continuous function which satisfies (2.88) and (2.89), by taking* z_0 *in place of* y_0, *for any* $x \in \mathbb{R}$. *Let* x *be an arbitrary real number. As the restrictions* $y_0\big|_{I_{n(x)}} (= y_{n(x)})$ *and* $z_0\big|_{I_{n(x)}}$ *both satisfy (2.88) as well as (2.89) for all* $x \in I_{n(x)}$, *the uniqueness of* $y_{n(x)} = y_0\big|_{I_{n(x)}}$ *implies that*

$$y_0(x) = y_0\big|_{I_{n(x)}}(x) = z_0\big|_{I_{n(x)}}(x) = z_0(x).$$

Hence, the theorem is proved.

Theorem 31 *Given* $c \in \mathbb{R}$ *and* $r > 0$, *let* I *denote a closed ball of radius* r *and centered at* c, *that is,* $I = \{x \in \mathbb{R} : c - r \leq x \leq c + r\}$ *and let* $F : I \times \mathbb{R} \to \mathbb{R}$ *be a continuous function which satisfies a Lipschitz condition (2.85) for all* $x \in I$ *and* $y, z \in \mathbb{R}$, *where* L *is a constant with* $0 < Lr < 1$. *If a continuously differentiable function* $y : I \to \mathbb{R}$ *satisfies the differential inequality*

$$|y'(x) - F(x, y(x))| \leq \varepsilon \tag{2.103}$$

for all $x \in I$ *and for some* $\varepsilon \geq 0$, *then there exists a unique continuous function* $y_0 : I \to \mathbb{R}$ *satisfying equation (2.88);* y_0 *is a solution to (2.84) and*

$$|y(x) - y_0(x)| \leq \frac{r}{1-Lr}\varepsilon \tag{2.104}$$

for any $x \in I$.

Proof 59 *Define a set* X *of all continuous functions* $f : I \to \mathbb{R}$ *by*

$$X = \{f : I \to \mathbb{R} : f \text{ is continuous}\}$$

and also introduce a generalized metric on X *as given below:*

$$d(f, g) = \inf\{C \in [0, \infty] : |f(x) - g(x)| \leq C, \forall x \in I\}.$$

By Theorem 29, it can be proved that (X, d) *is a generalized complete metric space.*

Consider, an operator $\Lambda : X \to X$ *defined by*

$$(\Lambda f)(x) = y(c) + \int_c^x F(\tau, f(\tau))d\tau, \; x \in I, \; \forall f \in X. \tag{2.105}$$

Assume that Λ *is strictly contractive on* X. *For each* $f, g \in X$, *suppose* $C_{fg} \in [0, \infty]$ *be an arbitrary constant such that* $d(f,g) \leq C_{fg}$. *Let's assume that*

$$|f(x) - g(x)| \leq C_{fg} \ \forall x \in I. \tag{2.106}$$

By (2.85), (2.105) *and* (2.106) *it follows that*

$$\begin{aligned}
|(\Lambda f)(x) - (\Lambda g)(x)| &= \left| \int_c^x \{F(\tau, f(\tau)) - F(\tau, g(\tau))\} d\tau \right| \\
&\leq \left| \int_c^x |F(\tau, f(\tau)) - F(\tau, g(\tau))| d\tau \right| \\
&\leq L \left| \int_c^x |f(\tau) - g(\tau)| d\tau \right| \\
&\leq LrC_{fg}, \ \forall x \in I
\end{aligned}$$

implies that $d(\Lambda f, \Lambda g) \leq LrC_{fg}$. *Therefore,* $d(\Lambda f, \Lambda g) \leq Lrd(f,g)$, $\forall f, g \in X$, *where* $0 < Lr < 1$. *By Theorem 29, for an arbitrary* $g_0 \in X$, *we can prove that*

$$d(\Lambda g_0, g_0) < \infty.$$

Hence, by Theorem 28(a), there exists a continuous function $y_0 : I \to \mathbb{R}$ *such that* $\Lambda^n g_0 \to y_0$ *in* (X, d) *as* $n \to \infty$ *and* $\Lambda y_0 = y_0$, *and hence,* y_0 *satisfies equation* (2.88), *for all* $x \in I$.

Let $g \in X$ *be arbitrary. As* g *and* g_0 *are bounded on a compact interval* I, *so there exists a constant* $C > 0$ *such that*

$$|g_0(x) - g(x)| \leq C, \ \forall x \in I.$$

Hence, $d(g_0, g) < \infty$ *for each* $g \in X$ *and* $\{g \in X | d(g_0, g) < \infty\} = X$. *By Theorem 28(b), we conclude that* y_0 *is the unique continuous function by using the property of* (2.88). (2.103) *implies that*

$$-\varepsilon \leq y'(x) - F(x, y(x)) \leq \varepsilon, \ \forall x \in I. \tag{2.107}$$

Integrating (2.107) *from* c *to* x, *we get*

$$|(\Lambda y)(x) - y(x)| \leq \varepsilon r, \ \forall x \in I$$

which implies that $d(\Lambda y, y) \leq \varepsilon r$. *Ultimately, from Theorem 28(c) we obtain*

$$d(y, y_0) \leq \frac{d(\Lambda y, y)}{1 - Lr} \leq \frac{r\varepsilon}{1 - Lr},$$

which is (2.104), *for all* $x \in I$.

Example 4 *Choose positive constants* K *and* L *such that* $KL < 1$. *Let* $I = [0, 2K - \varepsilon]$ *be a closed interval for a positive number* $\varepsilon < 2K$. *For a given* *polynomial* $p(x)$, *assume that a continuously differentiable function* $y : I \to \mathbb{R}$ *satisfies*

$$|y'(x) - Ly(x) - p(x)| \le x + \varepsilon, \ \forall x \in I.$$

Set $F(x, y) = Ly + p(x)$ *and* $\varphi(x) = x + \varepsilon$. *Hence, above inequality has the identical form of* (2.86). *So, we obtain*

$$\left| \int_0^x \varphi(\tau) d\tau \right| = \frac{1}{2}x^2 + \varepsilon x, \forall x \in I. \tag{2.108}$$

By (2.87), $\left| \int_0^x \varphi(\tau) d\tau \right| \le K\varphi(x)$. *Hence,* (2.108) *can be written as*

$$\left| \int_0^x \varphi(\tau) d\tau \right| = \frac{1}{2}x^2 + \varepsilon x \le K\varphi(x), \forall x \in I.$$

By Theorem 29, there exists a unique continuous function $y_0 : I \to \mathbb{R}$ *such that*

$$y_0(x) = y(0) + \int_0^x \{Ly_0(\tau) + p(\tau)\} d\tau$$

and

$$|y(x) - y_0(x)| \le \frac{K}{1 - KL}(x + \varepsilon)$$

for all $x \in I$.

Example 5 *Let* a *be a constant greater than* 1 *and choose a constant* L *with* $0 < L < lna$. *Given an interval* $I = [0, \infty)$ *and a polynomial* $p(x)$, *suppose* $y : I \to \mathbb{R}$ *is a continuously differentiable function satisfying*

$$|y'(x) - Ly(x) - p(x)| \le a^x, \ \forall x \in I.$$

Setting $\varphi(x) = a^x$, *we obtain*

$$\left| \int_0^x \varphi(\tau) d\tau \right| \le \frac{1}{lna}\varphi(x)$$

for all $x \in I$. *By Theorem 30, there exists a unique continuous function* $y_0 : I \to \mathbb{R}$ *such that*

$$y_0(x) = y(0) + \int_0^x \{Ly_0(\tau) + p(\tau)\} d\tau$$

and

$$|y(x) - y_0(x)| \le \frac{1}{lna - L} a^x$$

for all $x \in I$.

Example 6 *Assume that r and L are positive constants such that $0 < Lr < 1$ and define a closed interval $I = \{x \in \mathbb{R} : c - r \leq x \leq c + r\}$ for some real number c. Assume that a continuously differentiable function $y : I \to \mathbb{R}$ satisfies*

$$|y'(x) - Ly(x) - p(x)| \leq \varepsilon$$

for each $x \in I$ and for some $\varepsilon \geq 0$, where $p(x)$ is a polynomial. Then by Theorem 31, there exists a unique continuous function $y_0(x) : I \to \mathbb{R}$ such that

$$y_0(x) = y(c) + \int_c^x \{Ly_0(\tau) + p(\tau)\} d\tau$$

and

$$|y(x) - y_0(x)| \leq \frac{r}{1 - Lr}\varepsilon$$

for all $x \in I$.

2.7 Notes

To the best of our knowledge, S. M. Jung, T. Miura and M. Obloza should be credited with their contribution works [1], [5], [17], [18], [19], [20], [24], [37], [38], [42] and [65] on the Hyers-Ulam stability of first order differential equations of both linear and nonlinear type. Section 2.6 is especially meant for fixed point approach of Hyers-Ulam stability.

Chapter 3

Stability of Second Order Linear Differential Equations

In this chapter, we present the Hyers-Ulam stability of the most basic linear second order differential equations of the form:

$$y'' + \alpha y' + \beta y = 0, \tag{3.1}$$

$$y'' + \alpha y' + \beta y = f(x), \tag{3.2}$$

$$y'' + \alpha(x)y = 0, \tag{3.3}$$

$$y'' + \beta(x)y = f(x), \tag{3.4}$$

$$y'' + p(x)y' + q(x)y + r(x) = 0 \tag{3.5}$$

and

$$y'' + p(x)y' + q(x)y = f(x), \tag{3.6}$$

where $\alpha(x), \beta(x) \in \mathbb{R}$, $p, q, r \in C(\mathbb{R}, \mathbb{R})$, $y \in C^2[a, b]$ and $f \in C[a, b]$, $-\infty < a < b < +\infty$. We can not ignore such equations as long as their association with the physical problems are concerned.

3.1 Hyers-Ulam Stability of $y'' + \alpha y' + \beta y = 0$

Theorem 32 *If the characteristic equation $\lambda^2 + \alpha\lambda + \beta = 0$ has two different positive roots, then (3.1) has the Hyers-Ulam stability.*

Proof 60 *Let $\epsilon > 0$ and $y \in C^2[a, b]$ be such that*

$$| y'' + \alpha y' + \beta y | \leq \epsilon.$$

We show that there exists a constant K independent of ϵ and y such that $|y(x) - u(x)| < K\epsilon$, for some $u \in C^2[a, b]$ satisfying $u'' + \alpha u' + \beta u = 0$. Let λ_1 and λ_2 be the roots of characteristic equation $\lambda^2 + \alpha\lambda + \beta = 0$. Define

$g(x) = y'(x) - \lambda_1 y(x)$. *Then* $g'(x) = y''(x) - \lambda_1 y'(x)$. *Now,*

$$\begin{aligned} |\, g'(x) - \lambda_2 g(x) \,| &= |\, y''(x) - \lambda_1 y'(x) - \lambda_2(y'(x) - \lambda_1 y(x)) \,| \\ &= |\, y''(x) - \lambda_1 y'(x) - \lambda_2 y'(x) + \lambda_1\lambda_2 y(x) \,| \\ &= |\, y''(x) + (-\lambda_1 - \lambda_2) y'(x) + \lambda_1\lambda_2 y(x) \,| \\ &= |\, y'' + \alpha y'(x) + \beta y(x) \,| \le \epsilon \end{aligned}$$

implies that

$$-\epsilon \le g'(x) - \lambda_2 g(x) \le \epsilon.$$

Therefore, for any $x \in [a, b]$

$$-\epsilon e^{-\lambda_2(x-a)} \le g'(x)e^{-\lambda_2(x-a)} - \lambda_2 g(x) e^{-\lambda_2(x-a)} \le \epsilon e^{-\lambda_2(x-a)},$$

that is,

$$-\epsilon e^{-\lambda_2(x-a)} \le \frac{d}{dx}\left[g(x)e^{-\lambda_2(x-a)}\right] \le \epsilon e^{-\lambda_2(x-a)}. \tag{3.7}$$

Integrating (3.7) *from* x *to* b, *we get*

$$-\int_x^b \epsilon e^{-\lambda_2(y-a)} dy \le \int_x^b \frac{d}{dy}\left[g(y)e^{-\lambda_2(y-a)}\right] dy \le \int_x^b \epsilon e^{-\lambda_2(y-a)} dy,$$

that is,

$$\left[-\epsilon\frac{e^{-\lambda_2(y-a)}}{-\lambda_2}\right]_x^b \le \left[g(y)e^{-\lambda_2(y-a)}\right]_x^b \le \left[\epsilon\frac{e^{\lambda_2(y-a)}}{-\lambda_2}\right]_x^b. \tag{3.8}$$

Assume that $\lambda_2 > 1$. *If* $0 < \lambda_2 \le 1$, *then there exists* $M > 0$ *such that* $M\lambda_2 > 1$ *and the procedure can similarly be dealt with. Using* $\lambda_2 > 1$, (3.8) *can be written as*

$$[\epsilon e^{-\lambda_2(y-a)}]_x^b \le [g(y)e^{-\lambda_2(y-a)}]_x^b \le -[\epsilon e^{-\lambda_2(y-a)}]_x^b.$$

Hence,

$$\begin{aligned} \epsilon e^{-\lambda_2(b-a)} - \epsilon e^{-\lambda_2(x-a)} &\le g(b)e^{-\lambda_2(b-a)} - g(x)e^{-\lambda_2(x-a)} \\ &\le -\epsilon e^{-\lambda_2(b-a)} + \epsilon e^{-\lambda_2(x-a)}, \end{aligned}$$

that is,

$$-\epsilon e^{-\lambda_2(x-a)} \le (g(b) - \epsilon)e^{-\lambda_2(b-a)} - g(x)e^{-\lambda_2(x-a)} \le \epsilon e^{-\lambda_2(x-a)} - 2\epsilon e^{-\lambda_2(b-a)}$$

implies that

$$-\epsilon e^{-\lambda_2(x-a)} \le (g(b) - \epsilon)e^{-\lambda_2(b-a)} - g(x)e^{-\lambda_2(x-a)} \le \epsilon e^{-\lambda_2(x-a)}.$$

Consequently,

$$-\epsilon \leq (g(b) - \epsilon)e^{-\lambda_2(b-x)} - g(x) \leq \epsilon.$$

If $z(x) = (g(b) - \epsilon)e^{\lambda_2(x-b)}$, *then* $z'(x) = \lambda_2(g(b) - \epsilon)e^{\lambda_2(x-b)}$ *and*

$$z'(x) - \lambda_2 z(x) = \lambda_2(g(b) - \epsilon)e^{\lambda_2(x-b)} - \lambda_2(g(b) - \epsilon)e^{\lambda_2(x-b)} = 0. \tag{3.9}$$

So, $z(x)$ *satisfying* $z'(x) - \lambda_2 z(x) = 0$ *and* $| g(x) - z(x) | \leq \epsilon$. *Since* $g(x) = y'(x) - \lambda_1 y(x)$, *then*

$$-\epsilon \leq y'(x) - \lambda_1 y(x) - z(x) \leq \epsilon.$$

Therefore, for any $x \in [a, b]$

$$-\epsilon e^{-\lambda_1(x-a)} \leq y'(x)e^{-\lambda_1(x-a)} - \lambda_1 y(x)e^{-\lambda_1(x-a)} - z(x)e^{-\lambda_1(x-a)} \leq \epsilon e^{-\lambda_1(x-a)}.$$

Consequently,

$$-\epsilon e^{-\lambda_1(x-a)} \leq \frac{d}{dx}[y(x)e^{-\lambda_1(x-a)}] - z(x)e^{-\lambda_1(x-a)} \leq \epsilon e^{-\lambda_1(x-a)}. \tag{3.10}$$

Integrating (3.10) *from* x *to* b, *we get*

$$\begin{aligned} - \int_x^b \epsilon e^{-\lambda_1(s-a)} du \leq & \int_x^b \frac{d}{ds}[y(s)e^{-\lambda_1(s-a)}]ds - \int_x^b z(s)e^{-\lambda_1(s-a)} ds \\ & \leq \epsilon e^{-\lambda_1(s-a)} du, \end{aligned}$$

that is,

$$\begin{aligned} \left[-\epsilon \frac{e^{\lambda_1(s-a)}}{-\lambda_1}\right]_x^b \leq & \left[y(s)e^{-\lambda_1(s-a)}\right]_x^b - \int_x^b z(s)e^{-\lambda_1(s-a)} du \\ & \leq \left[\epsilon \frac{e^{-\lambda_1(s-a)}}{-\lambda_1}\right]_x^b. \end{aligned} \tag{3.11}$$

Using $\lambda_1 > 1$, (3.11) *can be written as*

$$\left[\epsilon e^{-\lambda_1(s-a)}\right]_x^b \leq \left[y(s)e^{-\lambda_1(s-a)}\right]_x^b - \int_x^b z(s)e^{-\lambda_1(s-a)} ds \leq \left[-\epsilon e^{-\lambda_1(s-a)}\right]_x^b,$$

that is,

$$\begin{aligned} \epsilon e^{-\lambda_1(b-a)} - \epsilon e^{-\lambda_1(x-a)} \leq & \left[y(b)e^{-\lambda_1(b-a)} - y(x)e^{-\lambda_1(x-a)}\right] \\ & - \int_x^b z(s)e^{-\lambda_1(s-a)} ds \\ & \leq -\epsilon e^{-\lambda_1(b-a)} + \epsilon e^{-\lambda_1(x-a)}. \end{aligned}$$

Therefore,

$$-\epsilon e^{-\lambda_1(x-a)} \le (y(b)-\epsilon)e^{-\lambda_1(b-a)} - y(x)e^{-\lambda_1(x-a)} - \int_x^b z(s)e^{-\lambda_1(s-a)}ds$$
$$\le \epsilon e^{-\lambda_1(x-a)}$$

implies that

$$-\epsilon \le (y(b)-\epsilon)e^{\lambda_1(x-b)} - y(x) - e^{\lambda_1(x-a)}\int_x^b z(s)e^{-\lambda_1(s-a)}ds \le \epsilon.$$

If we choose, $u(x) = (y(b)-\epsilon)e^{\lambda_1(x-b)} - e^{\lambda_1(x-a)}\int_x^b z(s)e^{-\lambda_1(s-a)}ds$, *then* $u'(x) = \lambda_1(y(b)-\epsilon)e^{\lambda_1(x-b)} - \lambda_1 e^{\lambda_1(x-a)}\int_x^b z(s)e^{-\lambda_1(s-a)}ds + z(x)$ *and*

$$u'(x) - \lambda_1 u(x) - z(x) = \lambda_1(y(b)-\epsilon)e^{\lambda_1(x-b)} - \lambda_1 e^{\lambda_1(x-a)}\int_x^b z(s)e^{-\lambda_1(s-a)}ds$$
$$+ z(x) - \lambda_1(y(b)-\epsilon)e^{\lambda_1(x-b)} + \lambda_1 e^{\lambda_1(x-a)}\int_x^b z(s)e^{-\lambda_1(s-a)} - z(x) = 0.$$

Hence, $z(x) = u'(x) - \lambda_1 u(x)$ *and* $z'(x) = u''(x) - \lambda_1 u'(x)$, $\mid y(x) - u(x) \mid \le \epsilon$. *Using this fact in* (3.9), *we get*

$$u''(x) - \lambda_1 u'(x) - \lambda_2\{u'(x) - \lambda_1 u(x)\} = 0,$$

that is,

$$u''(x) + (-\lambda_1 - \lambda_2)u'(x) + \lambda_1\lambda_2 u(x) = 0.$$

As a result,

$$u''(x) + \alpha u'(x) + \beta u(x) = 0.$$

This completes the proof of the theorem .

Theorem 33 *Assume that the characteristic equation* $\lambda^2 + \alpha\lambda + \beta = 0$ *has two different positive roots. Then for every* $\epsilon > 0$, $f \in C[a,b]$, $y \in C^2[a,b]$, *if*

$$\mid y'' + \alpha y' + \beta y - f(x) \mid \le \epsilon,$$

there exists some $u \in C^2[a,b]$ *and* $K > 0$ *satisfying*

$$u'' + \alpha u' + \beta u = f(x),$$

such that $\mid y(x) - u(x) \mid < K\epsilon$.

Proof 61 *Let* $\epsilon > 0$, $f \in C[a,b]$ *and* $y \in C^2[a,b]$ *be such that*

$$\mid y'' + \alpha y' + \beta y - f(x) \mid \le \epsilon.$$

Let λ_1 *and* λ_2 *be the two roots of the characteristic equation* $\lambda^2 + \alpha\lambda + \beta = 0$. *We may assume that* $\lambda_1 > 1$ *and* $\lambda_2 > 1$. *Define* $g(x) = y'(x) - \lambda_1 y(x)$. *Then* $g'(x) = y''(x) - \lambda_1 y'(x)$. *Now,*

$$\begin{aligned} \mid g'(x) - \lambda_2 g(x) - f(x) \mid &= \mid y''(x) - \lambda_1 y'(x) - \lambda_2(y'(x) - \lambda_1 y(x)) - f(x) \mid \\ &= \mid y''(x) - \lambda_1 y'(x) - \lambda_2 y'(x) + \lambda_1\lambda_2 y(x) - f(x) \mid \\ &= \mid y''(x) + (-\lambda_1 - \lambda_2)y'(x) + \lambda_1\lambda_2 y(x) - f(x) \mid \\ &= \mid y''(x) + \alpha y'(x) + \beta y(x) - f(x) \mid \leq \epsilon \end{aligned}$$

implies that

$$\mid g'(x) - \lambda_2 g(x) - f(x) \mid \leq \epsilon.$$

Therefore, for any $x \in [a, b]$

$$-\epsilon e^{-\lambda_2(x-a)} \leq g'(x)e^{-\lambda_2(x-a)} - \lambda_2 g(x)e^{-\lambda_2(x-a)} - f(x)e^{-\lambda_2(x-a)} \leq \epsilon e^{-\lambda_2(x-a)},$$

that is,

$$-\epsilon e^{-\lambda_2(x-a)} \leq \frac{d}{dx}\left[g(x)e^{-\lambda_2(x-a)}\right] - f(x)e^{-\lambda_2(x-a)} \leq \epsilon e^{-\lambda_2(x-a)}. \quad (3.12)$$

Integrating (3.12) *from* x *to* b, *we get*

$$\begin{aligned} -\int_x^b \epsilon e^{-\lambda_2(s-a)} ds &\leq \int_x^b \frac{d}{ds}\left[g(s)e^{-\lambda_2(s-a)}\right] ds - \int_x^b f(s)e^{-\lambda_2(s-a)} ds \\ &\leq \int_x^b \epsilon e^{-\lambda_2(s-a)} ds, \end{aligned}$$

that is,

$$\begin{aligned} \left[-\epsilon \frac{e^{-\lambda_2(s-a)}}{-\lambda_2}\right]_x^b &\leq \left[g(s)e^{-\lambda_2(s-a)}\right]_x^b - \int_x^b f(s)\frac{e^{-\lambda_2(s-a)}}{-\lambda_2} ds \\ &\leq \left[\epsilon \frac{e^{\lambda_2(s-a)}}{-\lambda_2}\right]_x^b. \end{aligned} \quad (3.13)$$

Using $\lambda_2 > 1$, (3.13) *becomes*

$$\begin{aligned} [\epsilon e^{-\lambda_2(s-a)}]_x^b &\leq [g(s)e^{-\lambda_2(s-a)}]_x^b - \int_x^b f(s)e^{-\lambda_2(s-a)} ds \\ &\leq -[\epsilon e^{-\lambda_2(s-a)}]_x^b. \end{aligned}$$

Hence,

$$\begin{aligned} \epsilon e^{-\lambda_2(b-a)} - \epsilon e^{-\lambda_2(x-a)} &\leq g(b)e^{-\lambda_2(b-a)} - g(x)e^{-\lambda_2(x-a)} - \int_x^b f(s)e^{-\lambda_2(s-a)} ds \\ &\leq -\epsilon e^{-\lambda_2(b-a)} + \epsilon e^{-\lambda_2(x-a)}, \end{aligned}$$

that is,

$$-\epsilon e^{-\lambda_2(x-a)} \leq (g(b)-\epsilon)e^{-\lambda_2(b-a)} - g(x)e^{-\lambda_2(x-a)} - \int_x^b f(s)e^{-\lambda_2(s-a)}ds$$
$$\leq \epsilon e^{-\lambda_2(x-a)}.$$

Consequently,

$$-\epsilon \leq (g(b)-\epsilon)e^{-\lambda_2(b-x)} - g(x) - e^{\lambda_2(x-a)}\int_x^b f(s)e^{-\lambda_2(s-a)}ds \leq \epsilon.$$

If we choose $z(x) = (g(b)-\epsilon)e^{\lambda_2(x-b)} - e^{\lambda_2(x-a)}\int_x^b f(s)e^{-\lambda_2(s-a)}ds$, *then* $z'(x) = \lambda_2(g(b)-\epsilon)e^{\lambda_2(x-b)} - \lambda_2 e^{\lambda_2(x-a)}\int_x^b f(s)e^{-\lambda_2(s-a)}ds + f(x)$ *and*

$$z'(x) - \lambda_2 z(x) - f(x) = \lambda_2(g(b)-\epsilon)e^{\lambda_2(x-b)} - \lambda_2 e^{\lambda_2(x-a)}$$
$$\times \int_x^b f(s)e^{-\lambda_2(s-a)}ds + f(x) - \lambda_2(g(b)-\epsilon)e^{\lambda_2(x-b)} + \lambda_2 e^{\lambda_2(x-a)}$$
$$\times \int_x^b f(s)e^{-\lambda_2(s-a)}ds - f(x) = 0. \quad (3.14)$$

So, $z(x)$ *satisfing* $z'(x) - \lambda_2 z(x) - f(x) = 0$, *and* $\mid g(x) - z(x) \mid \leq \epsilon$. *Since* $g(x) = y'(x) - \lambda_1 y(x)$, *then*

$$-\epsilon \leq y'(x) - \lambda_1 y(x) - z(x) \leq \epsilon.$$

Therefore, for any $x \in [a,b]$

$$-\epsilon e^{-\lambda_1(x-a)} \leq y'(x)e^{-\lambda_1(x-a)} - \lambda_1 y(x)e^{-\lambda_1(x-a)} - z(x)e^{-\lambda_1(x-a)} \leq \epsilon e^{-\lambda_1(x-a)}.$$

Consequently,

$$-\epsilon e^{-\lambda_1(x-a)} \leq \frac{d}{dx}[y(x)e^{-\lambda_1(x-a)}] - z(x)e^{-\lambda_1(x-a)} \leq \epsilon e^{-\lambda_1(x-a)}. \quad (3.15)$$

Integrating (3.15) *from* x *to* b, *we get*

$$-\int_x^b \epsilon e^{-\lambda_1(s-a)}ds \leq \int_x^b \frac{d}{ds}[y(s)e^{-\lambda_1(s-a)}]ds - \int_x^b z(s)e^{-\lambda_1(s-a)}ds$$
$$\leq \epsilon e^{-\lambda_1(s-a)}ds,$$

that is,

$$\left[-\epsilon\frac{e^{\lambda_1(s-a)}}{-\lambda_1}\right]_x^b \leq \left[y(s)e^{-\lambda_1(s-a)}\right]_x^b - \int_x^b z(s)e^{-\lambda_1(s-a)}ds$$
$$\leq \left[\epsilon\frac{e^{-\lambda_1(s-a)}}{-\lambda_1}\right]_x^b. \quad (3.16)$$

Using $\lambda_1 > 1$, (3.16) *can be written as*

$$\left[\epsilon e^{-\lambda_1(s-a)}\right]_x^b \le \left[y(s)e^{-\lambda_1(s-a)}\right]_x^b - \int_x^b z(s)e^{-\lambda_1(s-a)}ds \le \left[-\epsilon e^{-\lambda_1(s-a)}\right]_x^b,$$

that is,

$$\epsilon e^{-\lambda_1(b-a)} - \epsilon e^{-\lambda_1(x-a)} \le \left[y(b)e^{-\lambda_1(b-a)} - y(x)e^{-\lambda_1(x-a)}\right]$$
$$- \int_x^b z(s)e^{-\lambda_1(s-a)}ds \le -\epsilon e^{-\lambda_1(b-a)} + \epsilon e^{-\lambda_1(x-a)}.$$

Therefore,

$$-\epsilon e^{-\lambda_1(x-a)} \le (y(b)-\epsilon)e^{-\lambda_1(b-a)} - y(x)e^{-\lambda_1(x-a)} - \int_x^b z(s)e^{-\lambda_1(s-a)}ds$$
$$\le \epsilon e^{-\lambda_1(x-a)}$$

implies that

$$-\epsilon \le (y(b)-\epsilon)e^{\lambda_1(x-b)} - y(x) - e^{\lambda_1(x-a)}\int_x^b z(s)e^{-\lambda_1(s-a)}ds \le \epsilon.$$

If we choose $u(x) = (y(b)-\epsilon)e^{\lambda_1(x-b)} - e^{\lambda_1(x-a)}\int_x^b z(s)e^{-\lambda_1(s-a)}ds$, *then* $u'(x) = \lambda_1(y(b)-\epsilon)e^{\lambda_1(x-b)} - \lambda_1 e^{\lambda_1(x-a)}\int_x^b z(s)e^{-\lambda_1(s-a)}ds + z(x)$, *and*

$$u'(x) - \lambda_1 u(x) - z(x) = \lambda_1(y(b)-\epsilon)e^{\lambda_1(x-b)} - \lambda_1 e^{\lambda_1(x-a)}\int_x^b z(s)e^{-\lambda_1(s-a)}ds$$
$$+z(x) - \lambda_1(y(b)-\epsilon)e^{\lambda_1(x-b)} + \lambda_1 e^{\lambda_1(x-a)}\int_x^b z(s)e^{-\lambda_1(s-a)} - z(x) = 0.$$

Hence, $z(x) = u'(x) - \lambda_1 u(x)$ *and* $z'(x) = u''(x) - \lambda_1 u'(x)$, $\mid y(x) - u(x) \mid \le \epsilon$. *Using this fact in* (3.14), *we get*

$$u''(x) - \lambda_1 u'(x) - \lambda_2(u'(x) - \lambda_1 u(x)) - f(x) = 0,$$

that is,

$$u''(x) + (-\lambda_1 - \lambda_2)u'(x) + \lambda_1\lambda_2 u(x) - f(x) = 0.$$

As a result,

$$u''(x) + \alpha u'(x) + \beta u(x) = f(x).$$

Thus the proof is complete.

3.2 Hyers-Ulam Stability of $y'' + \beta(x)y = 0$

In this section, Hyers-Ulam stability of (3.3) is established by using the boundary conditions $y(a) = 0 = y(b)$, $-\infty < a < b < \infty$ and initial conditions $y(a) = 0 = y'(a)$, $-\infty < a < \infty$.

Theorem 34 *If* $\max \mid \beta(x) \mid < \frac{8}{(b-a)^2}$. *Then (3.3) has the Hyers-Ulam stability with the boundary conditions* $y(a) = 0 = y(b)$.

Proof 62 *For every* $\epsilon > 0$, $y \in C^2[a,b]$, *let* $\mid y'' + \beta(x)y \mid \leq \epsilon$ *with boundary conditions* $y(a) = 0 = y(b)$. *Let* $M = \max\{\mid y(x) \mid : x \in [a,b]\}$. *Since* $y(a) = 0 = y(b)$, *then there exists* $x_0 \in (a,b)$, *such that* $|y(x_0)| = M$. *By Taylor's theorem,*

$$y(a) = y(x_0) + y'(x_0)(x_0 - a) + \frac{y''(\xi)}{2}(x_0 - a)^2,$$

$$y(b) = y(x_0) + y'(x_0)(x_0 - b) + \frac{y''(\eta)}{2}(x_0 - b)^2.$$

By Rolle's theorem $y'(x_0) = 0$ *and therefore,*

$$|y''(\xi)| = \frac{2M}{(x_0 - a)^2},$$

$$|y''(\eta)| = \frac{2M}{(x_0 - b)^2}.$$

If $x_0 \in (a, \frac{a+b}{2}]$, *then*

$$a < x_0 \leq \frac{a+b}{2}.$$

Therefore,

$$\left(\frac{a+b}{2} - a\right)^2 \geq (x_0 - a)^2$$

and thus

$$\frac{2M}{(x_0 - a)^2} \geq \frac{2M}{\frac{(b-a)^2}{4}} = \frac{8M}{(b-a)^2}.$$

If $x_0 \in [\frac{a+b}{2}, b)$,*then*

$$\frac{a+b}{2} \leq x_0 < b.$$

Therefore,

$$\left(\frac{a+b}{2} - b\right)^2 \geq (x_0 - b)^2$$

and so is

$$\frac{2M}{(x_0 - b)^2} \geq \frac{2M}{\frac{(b-a)^2}{4}} = \frac{8M}{(b-a)^2}.$$

For $\xi, \eta \in (a, b)$,

$$|y''(\xi)| \geq \frac{8M}{(b-a)^2},$$

$$|y''(\eta)| \geq \frac{8M}{(b-a)^2}.$$

Therefore, for all $\xi, \eta \in (a, b)$

$$\max |y''(x)| \geq \frac{8M}{(b-a)^2} = \frac{8}{(b-a)^2} \max |y(x)|,$$

that is,

$$\max |y(x)| \leq \frac{(b-a)^2}{8} \max |y''(x)|.$$

Thus,

$$\max |y(x)| \leq \frac{(b-a)^2}{8} [\max |y''(x)| - \max |\beta(x)| \max |y(x)| + \max |\beta(x)| \max |y(x)|].$$

Since

$$\max |y''(x) - \beta(x)y(x)| \geq \max |y''(x)| - \max |\beta(x)| \max |y(x)|,$$

then for $\max |y''(x) - \beta(x)y| \leq \max |y''(x) + \beta(x)y|$, *we find*

$$\begin{aligned} \max |y(x)| &\leq \frac{(b-a)^2}{8} [\max |y''(x) + \beta(x)y| + \max |\beta(x)| \max |\ y(x)|] \\ &\leq \frac{(b-a)^2}{8} \varepsilon + \frac{(b-a)^2}{8} [\max |\beta(x)| \max |\ y(x)|]. \end{aligned}$$

Let $\eta = \frac{(b-a)^2 \max |\beta(x)|}{8}$, $K = \frac{(b-a)^2}{(8(1-\eta))}$. *Then* $\max |y(x)| \leq K\epsilon$, *that is,* $|y(x)| \leq \max |y(x)| \leq K\epsilon$ *implies that* $|y(x)| \leq K\epsilon$ *and hence by Definition* 8, $|y(x) - 0| \leq K\epsilon$. *Therefore,* $z_0(x) = 0$ *is a solution of* $y'' - \beta(x)y = 0$ *with* $z_0(a) = 0 = z_0(b)$ *and satisfying the relation* $|y - z_0| \leq K\epsilon$. *Hence* (3.3) *has the Hyers-Ulam stability.*

Theorem 35 *If* $\max |\beta(x)| < \frac{2}{(b-a)^2}$. *Then* (3.3) *has the Hyers-Ulam stability with initial conditions* $y(a) = 0 = y'(a)$.

Proof 63 *For every* $\epsilon > 0$ *and* $y \in C^2[a, b]$, *let's assume that* $|y'' + \beta(x)y| \leq \epsilon$ *with initial conditions* $y(a) = 0 = y'(a)$. *By Taylor's theorem,*

$$y(x) = y(a) + y'(a)(x-a) + \frac{y''(\xi)}{2}(x-a)^2.$$

So,

$$|y(x)| = |\frac{y''(\xi)}{2}(x-a)^2| \leq \max |y''(x) \frac{(b-a)^2}{2}|$$

implies that

$$\max|y(x)| \leq \max|y''(x)\frac{(b-a)^2}{2}|.$$

Therefore,

$$\max|y(x)| \leq \frac{(b-a)^2}{2}[\max|y''(x)| - \max|\beta(x)|\max|y(x)| \\ + \max|\beta(x)|\max|y(x)|].$$

Since,

$$\max|y''(x) - \beta(x)y(x)| \geq \max|y''(x)| - \max|\beta(x)|\max|y(x)|,$$

then using $\max|y''(x) - \beta(x)y| \leq \max|y''(x) + \beta(x)y|$, *we obtain*

$$\max|y(x)| \leq \frac{(b-a)^2}{2}[\max|y''(x) + \beta(x)y| + \max|\beta(x)|\max|y(x)|] \\ \leq \frac{(b-a)^2}{2}\epsilon + \frac{(b-a)^2}{2}[\max|\beta(x)|\max|y(x)|].$$

Let $\eta = \frac{(b-a)^2 \max|\beta(x)|}{2}$, $K = \frac{(b-a)^2}{(2(1-\eta))}$. *Then* $\max|y(x)| \leq K\epsilon$, *that is,* $|y(x)| \leq \max|y(x)| \leq K\epsilon$, *implies that* $|y(x)| \leq K\epsilon$ *and hence by definition,* $|y(x) - 0| \leq K\epsilon$. *Therefore,* $z_0(x) = 0$ *is a solution of* $y'' - \beta(x)y = 0$ *with* $z_0(a) = 0 = z_0'(a)$ *and satisfying the relation* $|y - z_0| \leq K\epsilon$. *Hence* (3.3) *has the Hyers-Ulam stability.*

3.3 Hyers-Ulam Stability of $y'' + \beta(x)y = f(x)$

In this section, Hyers-Ulam stability of (3.4) is established by using the initial conditions $y(a) = 0 = y'(a)$.

Theorem 36 *Suppose* $|\beta(x)| < M$, *where* $M = \frac{2}{(b-a)^2}$, $\varphi : [a, b] \to [0, \infty)$ *in an increasing function. The equation* (3.4) *has the Generalized Hyers-Ulam stability if for* $\theta_\varphi \in C(\mathbb{R}_+, \mathbb{R}_+)$ *and for each approximate solution* $y \in C^2[a, b]$ *of* (3.4) *satisfying*

$$|y'' - \beta(x)y - f(x)| \leq \varphi(x), \tag{3.17}$$

there exists a solution $z_0 \in C^2[a, b]$ *of* (3.4) *with condition* $y(a) = 0 = y'(a)$ *such that*

$$|y(x) - z_0(x)| \leq \theta_\varphi(x)$$

Proof 64 *Given that* $|\beta(x)| < M$, *where* $M = \frac{2}{(b-a)^2}$, *from the inequality* (3.17), *we have*

$$\varphi(x) \leq y'' - \beta(x)y - f(x) \leq \varphi(x) \leq \varphi(x). \tag{3.18}$$

Integrating (3.18) *from* a *to* x,

$$\int_a^x \varphi(t)dt \leq y'(x) - \int_a^x \beta(t)y(t)dt - \int_a^x f(t)dt \leq \int_a^x \varphi(t)dt. \qquad (3.19)$$

Again integrating (3.19) *from* a *to* r,

$$\int_a^r \int_a^x \varphi(t)dtdr \leq y(x) - \int_a^r \int_a^x \beta(t)y(t)dtdr - \int_a^r \int_a^x f(t)dtdr \leq \int_a^r \int_a^x \varphi(t)dtdr.$$

By using Replacement Lemma, we obtain

$$\int_a^x (x-t)\varphi(t)dt \leq y(x) - \int_a^x (x-t)\beta(t)y(t)dt$$
$$- \int_a^x (x-t)f(t)dt \leq \int_a^x (x-t)\varphi(t)dt.$$

Hence,

$$\left| y(x) - \int_a^x (x-t)\beta(t)y(t)dt - \int_a^x (x-t)f(t)dt \right| \leq \int_a^x (x-t)\varphi(t)dt,$$

that is,

$$\left| y(x) - \int_a^x (x-t)\{\beta(t)y(t)dt + f(t)dt\} \right| \leq \int_a^x (x-t)\varphi(t)dt.$$

Let us consider $z_0(x) = \int_a^x (x-t)\{\beta(t)z_0(t)dt + f(t)dt\}$, *then*

$$|y(x) - z_0(x)| = \Bigg| y(x) - \int_a^x (x-t)\{\beta(t)y(t)dt + f(t)dt\}$$
$$+ \int_a^x (x-t)\{\beta(t)y(t)dt + f(t)dt\} - \int_a^x (x-t)\{\beta(t)z_0(t)dt + f(t)dt\} \Bigg|$$
$$\leq \left| y(x) - \int_a^x (x-t)\{\beta(t)y(t)dt + f(t)dt\} \right|$$
$$+ \Bigg| \int_a^x (x-t)\{\beta(t)y(t)dt + f(t)dt\}$$
$$- \int_a^x (x-t)\{\beta(t)z_0(t)dt + f(t)dt\} \Bigg|,$$

that is,

$$|y(x) - z_0(x)| \leq \int_a^x (x-t)\varphi(t)dt + \left| \int_a^x (x-t)\beta(t)\{y(t) - z_0(t)\}dt \right|$$

implies that

$$|y(x) - z_0(x)| \leq \int_a^x (x-t)\varphi(t)dt + \mid \beta(t) \mid \int_a^x (x-t) \mid y(t) - z_0(t) \mid dt$$

that is,

$$|y(x) - z_0(x)| \leq \int_a^x (x-t)\varphi(t)dt + M\int_a^x (x-t) \mid y(t) - z_0(t) \mid dt.$$

Applying Gronwall inequality, we have

$$|y(x) - z_0(x)| \leq \int_a^x (x-t)\varphi(t)dt e^{M\int_a^x (x-t)dt}$$

that is,

$$|y(x) - z_0(x)| \leq \int_a^x (x-t)\varphi(t)dt e^{\frac{M(x-a)^2}{2}}.$$

Therefore,

$$|y(x) - z_0(x)| \leq c\int_a^x (x-t)\varphi(t)dt,$$

where $c = e^{\left(\frac{x-a}{b-a}\right)^2}$.

Remark 3 *This is to be noted that as* $x \to b$*, the above system considered is the Hyers-Ulam stable.*

3.4 Hyers-Ulam Stability of $y'' + p(x)y' + q(x)y + r(x) = 0$

In this section, Hyers-Ulam stability of (3.5) is presented by means of Riccati differential equation.

Theorem 37 *Let* $p(x), q(x)$ *and* $r(x)$ *be real continuous functions on* $I = (a,b)$ *such that* $p(x) \neq 0$, $r(x) \neq 0$, $q(x) \neq 0$. *If a twice continuously differentiable function* $y : I \to \mathbb{R}$ *satisfies the differential inequality*

$$\mid y'' + p(x)y' + q(x)y + r(x) \mid \leq \varepsilon$$

for all $t \in I$ *and for* $\varepsilon > 0$ *and the Riccati equation* $u'(x) + p(x)u(x) - u^2(x) = q(x)$ *has a particular solution* $c(x)$*, then there exists a solution* $v : I \to \mathbb{R}$ *of* $y'' + p(x)y' + q(x)y + r(x) = 0$ *such that*

$$|y(x) - v(x)| \leq K\varepsilon,$$

where $K > 0$ *is a constant,* $p(x) - c(x) \geq 1$ *for all* $x \in I$.

Proof 65 *Let* $\varepsilon > 0$ *and* $y : I \to \mathbb{R}$ *be a continuously differentiable function satisfy the differential inequality*

$$\mid y'' + p(x)y' + q(x)y + r(x) \mid \leq \varepsilon.$$

We show that there exists a constant K independent of ε and y such that $|y(x)-v(x)| < K\varepsilon$ for some $v \in C^2(I)$ satisfying $v''+p(x)v'+q(x)v+r(x) = 0$. Let $c(x)$ be a particular solution of the Riccati equation

$$u'(x) + p(x)u(x) - u^2(x) = q(x).$$

Then $c'(x) + p(x)c(x) - c^2(x) = q(x)$. Define $g(x) = y'(x) + c(x)y(x)$, then $g'(x) = y''(x) + c(x)y'(x) + c'(x)y(x)$ and let $d(x) = p(x) - c(x)$. Therefore, $q(x) = c'(x) + d(x)c(x)$. Now,

$$\begin{aligned} \mid g'(x) + d(x)g(x) + r(x) \mid &= \mid y''(x) + c(x)y'(x) + c'(x)y(x) + d(x)\{y'(x) \\ &\quad + c(x)y(x)\} + r(x) \mid \\ &= \mid y''(x) + \{c(x) + d(x)\}y'(x) \\ &\quad + \{c'(x) + d(x)c(x)\}y(x) + r(x) \mid \\ &= \mid y'' + p(x)y' + q(x)y + r(x) \mid \\ &\leq \epsilon. \end{aligned}$$

Using the same technique as in Section 2.4, we have

$$w(x) = e^{-\int_a^x d(s)ds}\left[(g(b) - \epsilon)e^{\int_a^b d(x)ds} + \int_x^b r(s)e^{\int_a^s d(t)dt}ds\right]$$

and

$$\begin{aligned} w'(x) = &-d(x)e^{-\int_a^x d(s)ds}(g(b) - \epsilon)e^{\int_a^b d(x)ds} - r(x) \\ &- d(x)e^{-\int_a^x d(s)ds}\int_x^b r(s)e^{\int_a^s d(t)dt}ds. \end{aligned}$$

Hence,

$$\begin{aligned} w'(x) + d(x)w(x) + r(x) &= -d(x)e^{-\int_a^x d(s)ds}(g(b) - \epsilon)e^{\int_a^b d(x)ds} - r(x) \\ &\quad - d(x)e^{-\int_a^x d(s)ds}\int_x^b r(s)e^{\int_a^s d(t)dt}ds \\ &\quad + d(x)e^{-\int_a^x d(s)ds}(g(b) - \epsilon)e^{\int_a^b d(x)ds} \\ &\quad + d(x)e^{-\int_a^x d(s)ds}\int_x^b r(s)e^{\int_a^s d(t)dt}ds + r(x) \\ &= 0. \end{aligned} \tag{3.20}$$

So, $w(x)$ satisfying $w'(x)+d(x)w(x)+r(x) = 0$, and $\mid g(x)-w(x) \mid \leq \epsilon$. Since $g(x) = y'(x) + c(x)y(x)$, then

$$-\epsilon \leq y'(x) + c(x)y(x) - w(x) \leq \epsilon.$$

Using the same technique as above, we get

$$v(x) = e^{-\int_a^x c(s)ds}\left[(y(b) - \epsilon)e^{\int_a^b c(x)ds} - \int_x^b w(s)e^{\int_a^s c(t)dt}ds\right]$$

and

$$v'(x) = -c(x)e^{-\int_a^x c(s)ds}(g(b)-\epsilon)e^{\int_a^b c(x)ds} + w(x)$$
$$+ c(x)e^{-\int_a^x c(s)ds}\int_x^b w(s)e^{\int_a^s c(t)dt}ds.$$

So,

$$v'(x) + c(x)v(x) - w(x) = 0.$$

Hence $w(x) = v'(x) + c(x)v(x)$ *and* $w'(x) = v''(x) + c'(x)v(x) + v'(x)c(x) + d(x)v'(x)$, $|y(x) - v(x)| \leq \epsilon$. *Using this fact in* (3.20), *we get*

$$v''(x) + c'(x)v(x) + v'(x)c(x) + d(x)v'(x) + d(x)c(x)v(x) + r(x) = 0$$

implies that

$$v''(x) + (c(x) + d(x))v'(x) + (c'(x) + d(x)c(x))v(x) + r(x) = 0,$$

that is,

$$v''(x) + p(x)v'(x) + q(x)v(x) + r(x) = 0.$$

This completes the proof of the theorem.

Theorem 38 *Let* $p(x), q(x)$ *and* $r(x)$ *be continuous real functions defined on the interval* $I = (a,b)$ *such that* $p(x) \neq 0$ *and* $y_0(x)$ *is a non-zero bounded particular solution* $p(x)y'' + q(x)y' + r(x)y = 0$. *If* $y : I \to \mathbb{R}$ *is a twice continuously differentiable function,which satisfies the differential inequality*

$$|p(x)y'' + q(x)y' + r(x)y| \leq \epsilon$$

for all $t \in I$ *and for some* $\epsilon > 0$, *then there exists a solution* $v : I \to \mathbb{R}$ *such that*

$$|y(x) - v(x)| \leq K\epsilon,$$

where $K > 0$ *is a constant and* v *satisfies* $p(x)v'' + q(x)v' + r(x)v = 0$.

Proof 66 *Let* $\epsilon > 0$ *and* $y : I \to \mathbb{R}$ *be a twice continuously differentiable function, satisfy the differential inequality*

$$|p(x)y'' + q(x)y' + r(x)y| \leq \epsilon.$$

Since $y_0(x)$ *is a particular solution of* $p(x)y'' + q(x)y' + r(x)y = 0$, *then* $p(x)y_0''(x) + q(x)y_0'(x) + r(x)y_0(x) = 0$. *Consider* $y(x) = y_0(x)\int_a^x z(s)ds$. *Then*

$$| \, p(x)y''(x) + q(x)y'(x) + r(x)y(x) \, |$$
$$= |p(x)y_0''(x)\int_a^x z(x)ds + 2p(x)y_0'(x)z(x) + p(x)y_0(x)z'(x)$$
$$+ q(x)y_0'(x)\int_a^x z(s)ds + q(x)y_0(x)z(x) + r(x)y_0(x)\int_a^x z(s)ds|$$

$$= | p(x)y_0(x)z'(x) + 2p(x)y_0'(x)z(x) + q(x)y_0(x)z(x)$$
$$+ \int_a^x z(x)ds \{p(x)y_0''(x) + q(x)y_0'(x) + r(x)y_0(x)\} |$$
$$= | p(x)y_0(x)z'(x) + [2p(x)y_0'(x) + q(x)y_0(x)]z(x) |$$
$$\leq \epsilon.$$

Using the same technique as in Section 2.4, we have

$$z_1(x) = e^{-\int_a^x \frac{2p(s)y_0'(s)+q(s)}{y_0(s)}ds}\left[(z(b) - \varepsilon)e^{\int_a^b \frac{2p(s)y_0'(s)+q(s)}{y_0(s)}ds}\right]$$

So, $z_1(x)$ *satisfying* $p(x)y_1(x)z_1'(x) + [2p(x)y_1'(x) + q(x)y_1(x)]z_1(x) = 0$ *and* $| z(x) - z_1(x) | \leq \varepsilon$ *implies that*

$$-\varepsilon \leq z(x) - z_1(x) \leq \varepsilon.$$

Using the same technique as above, we get

$$z_2(x) = \left(\frac{y(b)}{y_1(b)} - \varepsilon\right) - \int_x^b z_1(s)ds.$$

So, $z_2(x)$ *satisfies* $z_2'(x) - z_1(x) = 0$ *and* $| z(x) - z_2(x) | \leq \varepsilon$*. Consequently,*

$$| z(x) - z_2(x)y_0(x) | \leq \varepsilon.$$

If we define $v(x) = z_2(x)y_0(x)$*, then the above inequality is*

$$| y(x) - v(x) | \leq K\varepsilon,$$

where $K > 0$ *is a constant and* v *satisfies* $p(x)v'' + q(x)v' + r(x)v = 0$.

Corollary 21 *Let* $p(x), q(x)$ *and* $r(x)$ *be continuous real functions defined on the interval* $I = (a, b)$ *such that* $p(x) \neq 0$ *and* $r^2 + p(x)r + q(x) = 0$*. If* $y : I \to \mathbb{R}$ *is a twice continuously differentiable function, which satisfies the differential inequality*

$$| p(x)y'' + q(x)y' + r(x)y | \leq \varepsilon$$

for all $t \in I$ *and for some* $\varepsilon > 0$*, then there exists a solution* $v : I \to \mathbb{R}$ *such that*

$$| y(x) - v(x) | \leq K\varepsilon,$$

where $K > 0$ *is a constant and* v *satisfies* $p(x)v'' + q(x)v' + r(x)v = 0$.

Corollary 22 *Let* $p(x)$*,* $q(x)$*,* $r(x)$*, and* $s(x)$ *be continuous real functions defined on the interval* $I = (a, b)$ *such that* $p(x) \neq 0$ *and* $y_0(x)$ *is a non-zero bounded particular solution of* $p(x)y''' + q(x)y'' + r(x)y' + s(x)y = 0$*. If*

$y : I \to \mathbb{R}$ is a twice continuously differentiable function, which satisfies the differential inequality

$$| p(x)y''' + q(x)y'' + r(x)y' + s(x)y | \leq \varepsilon$$

for all $t \in I$ and for some $\varepsilon > 0$, then there exists a solution $v : I \to \mathbb{R}$ such that

$$| y(x) - v(x) | \leq K\varepsilon,$$

where $K > 0$ is a constant and v satisfies $p(x)v''' + q(x)v'' + r(x)v' + s(x)v = 0$.

3.5 Hyers-Ulam Stability of $y'' + p(x)y' + q(x)y = f(x)$

In this section, the generalized Hyers-Ulam stability of differential equation (3.6) is presented.

Theorem 39 *Let X be a complex Banach space. Assume that $p, q : I \to C$ and $f : I \to X$ are continuous functions and $y_1 : I \to X$ is a non-zero twice continuously differentiable function which satisfies the differential equation*

$$y_1'' + p(x)y_1' + q(x)y_1 = 0. \tag{3.21}$$

If a twice continuously differentiable function $y : I \to X$ satisfies

$$\|y'' + p(x)y' + q(x)y - f(x)\| \leq \varphi(x) \tag{3.22}$$

for all $x \in I$ and $\varphi : I \to (0, \infty)$ is a continuous function, then there exists a unique $x_0 \in X$ such that

$$\begin{aligned} &\|y(x) - y_1(x). \int_a^x e^{\int_a^s \left(\frac{2y_1'(u)}{y_1(u)} + p(u)\right) du} \\ &\qquad \cdot \left[x_0 + \int_a^s \frac{f(t)}{y_1(t)} e^{\int_a^t \left(\frac{2y_1'(u)}{y_1(u)} + p(u)\right) du} dt \right] ds + k\| \\ &\leq \|y_1(x)\|. \int_a^x e^{\int_a^s \left(\frac{2y_1'(u)}{y_1(u)} + p(u)\right) du} \\ &\qquad \left| \int_s^b e^{\int_a^t \left(\frac{2y_1'(u)}{y_1(u)} + p(u)\right) du} \cdot \frac{\varphi(t)}{\|y_1(t)\|} dt \right| ds, \end{aligned}$$

where $k = \frac{y(a)}{y_1(a)} \in \mathbb{C}$ and $e^{\Re \int_a^t (\frac{2y_1'(a)}{y_1(a)} + p(u))du} \varphi(t)$ is integrable.

Proof 67 *Assume that* $v(x) = \frac{y(x)}{y_1(x)}$. *Then (3.22) becomes*

$$\begin{aligned}
&\|(v(x)y_1(x))'' + p(x)(v(x)y_1(x))' + q(x)v(x)y_1(x) - f(x)\| \\
&\quad = \|v(x)''y_1(x) + v'(x)y_1'(x) + v'(x)y_1'(x) + p(x)v'(x)y_1(x) \\
&\quad + v(x)y_1''(x) + p(x)v(x)y_1'(x) + q(x)v(x)y_1(x) - f(x)\| \\
&\quad = \|v''(x)y_1(x) + 2v'(x)y_1'(x) + v(x)y_1''(x) + p(x)v'(x)y_1(x) \\
&\quad + p(x)v(x)y_1'(x) + q(x)v(x)y_1(x) - f(x)\| \\
&\quad = \|v''(x)y_1(x) + v'(x)(2y_1'(x) + p(x)y_1(x)) \\
&\quad + v(x)(y_1''(x) + p(x)y_1'(x) + q(x)y_1(x)) - f(x)\|,
\end{aligned}$$

that is,

$$\begin{aligned}
&\|v''(x)y_1(x) + v'(x)(2y_1'(x) + p(x)y_1'(x)) - f(x)\| \\
&= \left\| y_1(x)\left\{ v''(x) + v'(x)\left(\frac{2y_1'(x)}{y_1(x)} + p(x)\right) - \frac{f(x)}{y_1(x)} \right\} \right\| \\
&\le \varphi(x).
\end{aligned}$$

Therefore,

$$\left\| v''(x) + v'(x)\left(\frac{2y_1'(x)}{y_1(x)} + p(x)\right) - \frac{f(x)}{y_1(x)} \right\| \le \frac{\varphi(x)}{\| y_1(x) \|}.$$

Consider

$$z(s) = e^{\int_a^s \left(\frac{2y_1'(u)}{y_1(u)} + p(u)\right) du} \cdot \left(\frac{y(s)}{y_1(s)}\right)' - \int_a^s \left(\frac{f(t)}{y_1(t)} e^{\int_a^t \left(\frac{2y_1'(u)}{y_1(u)} + p(u)\right) du}\right) dt.$$

For all $l, s \in I$,

$$\begin{aligned}
\|z(s) - z(l)\| &= \left\| e^{\int_a^s \left(\frac{2y_1'(u)}{y_1(u)} + p(u)\right) du} \cdot \left(\frac{y(s)}{y_1(s)}\right)' - e^{\int_a^l \left(\frac{2y_1'(u)}{y_1(u)} + p(u)\right) du} \cdot \left(\frac{y(l)}{y_1(l)}\right)' \right. \\
&- \int_a^s \left(\frac{f(t)}{y_1(t)} e^{\int_a^t \left(\frac{2y_1'(u)}{y_1(u)} + p(u)\right) du}\right) dt + \left. \int_a^l \left(\frac{f(t)}{y_1(t)} e^{\int_a^t \left(\frac{2y_1'(u)}{y_1(u)} + p(u)\right) du}\right) dt \right\| \\
&= \left\| \int_l^s \frac{d}{dt}\left[e^{\int_a^t \left(\frac{2y_1'(u)}{y_1(u)} + p(u)\right) du} \cdot \left(\frac{y(t)}{y_1(t)}\right)' \right] dt \right. \\
&- \left. \int_l^s \left(\frac{f(t)}{y_1(t)} e^{\int_a^t \left(\frac{2y_1'(u)}{y_1(u)} + p(u)\right) du}\right) dt \right\| \\
&= \left\| \int_l^s \left(\frac{2y_1'(t)}{y_1(t)} + p(t)\right) . e^{\int_a^t \left(\frac{2y_1'(u)}{y_1(u)} + p(u)\right) du} \cdot \left(\frac{y(t)}{y_1(t)}\right)' dt \right.
\end{aligned}$$

$$+ e^{\int_a^t \left(\frac{2y_1'(u)}{y_1(u)}+p(u)\right)du} \cdot \left(\frac{y(t)}{y_1(t)}\right)'' dt - \int_l^s \left(\frac{f(t)}{y_1(t)} e^{\int_a^t \left(\frac{2y_1'(u)}{y_1(u)}+p(u)\right)du}\right) dt \Bigg\|$$

$$= \Bigg\| \int_l^s e^{\int_a^t \left(\frac{2y_1'(u)}{y_1(u)}+p(u)\right)du} \Bigg\{ \left(\frac{2y_1'(t)}{y_1(t)} + p(t)\right) \cdot \left(\frac{y(t)}{y_1(t)}\right)'$$

$$+ \left(\frac{y(t)}{y_1(t)}\right)'' - \frac{f(t)}{y_1(t)} \Bigg\} dt \Bigg\|$$

$$\leq \left| \int_l^s e^{\left(\int_a^t \left(\frac{2y_1'(u)}{y_1(u)}+p(u)\right)du\right)} \cdot \frac{\varphi(t)}{\| y_1(t) \|} dt \right| .$$

Since $e^{\int_a^t \left(\frac{2y_1'(u)}{y_1(u)}+p(u)\right)du} \cdot \varphi(t)$ *is integrable on* I, *then we can find* $l_0 \in I$, *such that* $\|z(s) - z(l)\| < \epsilon$ *for* $l, s \geq l_0$, *that is* $z(l)_{l \in I}$ *is a Cauchy net and* $x_0 \in X$ *such that* $z(l)$ *converges to* x_0 *as* $l \to b$. *Now,*

$$\Bigg\| y(x) - y_1(x) \cdot \int_a^x e^{-\int_a^s \left(\frac{2y_1'(u)}{y_1(u)}+p(u)\right)du}$$

$$\cdot \left[x_0 + \int_a^s \frac{f(t)}{y_1(t)} e^{\int_a^t \left(\frac{2y_1'(u)}{y_1(u)+p(u)}\right)du} dt \right] ds + k \Bigg\|$$

$$= \Bigg\| y_1(x). \left\{ \int_a^x e^{-\int_a^s \left(\frac{2y_1'(u)}{y_1(u)}+p(u)\right)du} .(z(s) - x_0) \right\} ds \Bigg\|$$

$$\leq \| y_1(x) \| \cdot \int_a^x \left\{ e^{-\Re \int_a^s \left(\frac{2y_1'(u)}{y_1(u)}+p(u)\right)du} . \| z(s) - z(l) \| \right\} ds$$

$$+ \| y_1(x) \| \cdot \int_a^x \left\{ e^{-\Re \int_a^s \left(\frac{2y_1'(u)}{y_1(u)}+p(u)\right)du} . \| z(l) - x_0 \| \right\} ds$$

$$\leq \| y_1(x) \| \cdot \int_a^x e^{-\Re \int_a^s \left(\frac{2y_1'(u)}{y_1(u)}+p(u)\right)du}$$

$$\cdot \left| \int_l^s e^{\Re \int_a^t \left(\frac{2y_1'(u)}{y_1(u)}+p(u)\right)du} \cdot \frac{\varphi(t)}{\|y_1(t)\|} dt \right| ds$$

$$+ \| y_1(x) \| \cdot \int_a^x \left(e^{-\Re \int_a^s \left\{\frac{2y_1'(u)}{y_1(u)}+p(u)\right\}du} . \| z(l) - x_0 \| \right) ds$$

$$= \|y_1(x)\| \cdot \int_a^x e^{-\Re \int_a^s \left(\frac{2y_1'(u)}{y_1(u)}+p(u)\right)du}$$

$$\cdot \left| \int_b^s e^{\Re \int_a^t \left(\frac{2y_1'(u)}{y_1(u)}+p(u)\right)du} \cdot \frac{\varphi(t)}{\|y_1(t)\|} dt \right| ds,$$

as $z(l) \to x_0$ and $l \to b$ implies that

$$\| y(x) - y_1(x). \int_a^x e^{\int_a^s \left(\frac{2y_1'(u)}{y_1(u)} + p(u)\right) du} \cdot \left[x_0 + \int_a^s \frac{f(t)}{y_1(t)} e^{\int_a^t \left(\frac{2y_1(u)}{y_1(u)} + p(u)\right) du} dt \right] ds + k \|$$
$$\leq \|y_1(x)\| \cdot \int_a^x e^{-\Re \int_a^s \left(\frac{2y_1'(u)}{y_1(u)} + p(u)\right) du} \cdot \left| \int_b^s e^{\Re \int_a^t \left(\frac{2y_1'(u)}{y_1(u)} + p(u)\right) du} \cdot \frac{\varphi(t)}{\|y_1(t)\|} dt \right| ds. \tag{3.23}$$

If,

$$y_0(x) = y_1(x) \cdot \int_a^x e^{-\int_a^s \left(\frac{2y_1'(u)}{y_1(u)} + p(u)\right) du} \cdot \left[x_0 + \int_a^s \frac{f(t)}{y_1(t)} e^{\int_a^t \left(\frac{2y_1'(u)}{y_1(u) + p(u)}\right) du} dt \right] ds + k,$$

then (3.23) becomes

$$\| y(x) - y_0(x) \| \leq \|y_1(x)\| \cdot \int_a^x e^{-\Re \int_a^s \left(\frac{2y_1'(u)}{y_1(u)} + p(u)\right) du} \cdot \left| \int_b^s e^{\Re \int_a^t \left(\frac{2y_1'(u)}{y_1(u)} + p(u)\right) du} \cdot \frac{\varphi(t)}{\|y_1(t)\|} dt \right| ds.$$

Now, it remains to show the uniqueness of x_0. Assume that $x_1, x_2 \in X$ satisfy the inequality (3.23) in place of x_0. Then

$$\Bigg\| y(x) - y_1(x). \int_a^x e^{-\int_a^s \left(\frac{2y_1'(u)}{y_1(u)} + p(u)\right) du} . \, x_1 ds$$
$$- y_1(x). \int_a^x \left(e^{-\int_a^s \left(\frac{2y_1'(u)}{y_1(u)} + p(u)\right) du} . \int_a^s \frac{f(t)}{y_1(t)} e^{\int_a^t \left(\frac{2y_1'(u)}{y_1(u)} + p(u)\right) du} dt \right) ds$$
$$- y(x) + y_1(x). \int_a^x e^{-\int_a^s \left(\frac{2y_1'(u)}{y_1(u)} + p(u)\right) du} . \, x_2 ds$$
$$+ y_1(x). \int_a^x \left(e^{-\int_a^s \left(\frac{2y_1'(u)}{y_1(u)} + p(u)\right) du} . \int_a^s \frac{f(t)}{y_1(t)} e^{\int_a^t \left(\frac{2y_1'(u)}{y_1(u)} + p(u)\right) du} dt \right) ds \Bigg\|$$
$$= \Bigg\| y_1(x). \int_a^x e^{-\int_a^s \left(\frac{2y_1'(u)}{y_1(u)} + p(u)\right) du} . \, x_1 ds$$

$$- y_1(x). \int_a^x e^{-\int_a^s \left(\frac{2y_1'(u)}{y_1(u)}+p(u)\right)du} .x_2 ds\Bigg\|$$

$$= \Bigg\| y_1(x). \left\{ \int_a^x \left(e^{-\int_a^s \left(\frac{2y_1'(u)}{y_1(u)}+p(u)\right)du} .(x_2 - x_1)\right) ds \right\} \Bigg\|$$

$$\leq \| y_1(x) \| . \int_a^x e^{-\Re \int_a^s \left(\frac{2y_1'(u)}{y_1(u)}+p(u)\right)du}$$

$$. \left| \int_s^b e^{\Re \int_a^t \left(\frac{2y_1'(u)}{y_1(u)}+p(u)\right)du} . \frac{\varphi(t)}{\| y_1(t) \|} dt \right| ds,$$

that is,

$$\| x_2 - x_1 \| \leq \frac{\| y_1(x) \| \int_a^x \left\{ e^{-\Re \int_a^s M du} . \mid \int_s^b e^{\Re \int_a^t M du} . \left(\frac{\varphi(t)}{\|y_1(t)\|}\right) dt \mid \right\} ds}{\| y_1(x) \| \int_a^x \left(e^{-\Re \int_a^s M du} \right) ds},$$

where $M = \left(\frac{2y_1'(u)}{y_1(u)} + p(u)\right)$. *Indeed,*

$$\left| \int_s^b e^{\Re \int_a^t \left(\frac{2y_1'(u)}{y_1(u)}+p(u)\right)du} . \frac{\varphi(t)}{\| y_1(t) \|} dt \right| \to 0$$

as $s \to b$ *implies that*

$$x_2 = x_1.$$

This completes the proof of the theorem.

Remark 4 [14] *It follows from Theorem* 39 *that*

$$y(x) = y_1(x) \cdot \int_a^x e^{\int_a^s \left(\frac{2y_1'(u)}{y_1(u)}+p(u)\right)du}$$

$$\cdot \left[c_1 + \int_a^s e^{\int_a^v \left(\frac{2y_1'(u)}{y_1(u)}+p(u)\right)du} dv \right] ds + c_2,$$

is the general solution of the equation (3.6), *where* c_1, c_2 *are arbitrary elements of* X *and* $y_1(x)$ *is a non-zero solution of the corresponding homogeneous equation* (3.6).

Remark 5 *If we replace* $\mathbb{C}$ *by* $\mathbb{R}$ *in the proof of Theorem 39 and if we assume that* p, q *are real-valued continuous functions,then we can see that Theorem 39 is true for a real Banach space* X.

Hence, every second order linear differential equation has the generalized Hyers-Ulam stability with the condition that there exists a solution of corresponding homogeneous equation or there exists a general solution in ordinary differential equations.

Example 7 *Consider the second order linear differential equation with constant coefficients*

$$y'' + by' + cy = f(x) \tag{3.24}$$

Its auxiliary equation is $m^2 + m + c = 0$, *where* $m = \frac{-b\pm\sqrt{b^2-4c}}{2}$. *Let* $b^2 - 4c \geq 0$, *and let* $f : I \to \mathbb{R}$, $\varphi : I \to [0, \infty)$ *be continuous functions. Assume that* $y : I \to \mathbb{R}$ *is twice continuously differential function satisfying the differential inequality*

$$\mid y'' + by' + cy - f(x) \mid \leq \varphi(x)$$

for all $x \in I$. *Since* $y_1(x) = e^{mx}$ *is a solution of* (3.24), *then using Theorem 39, there exists a solution* $y_0 : I \to \mathbb{R}$ *such that*

$$\begin{aligned} y_0(x) &= y_1(x) \cdot \int_a^x e^{-\Re \int_a^s \left(\frac{2y_1'(u)}{y_1(u)}+p(u)\right)du} \\ &\quad \cdot \left| x_0 + \int_b^s e^{\Re \int_a^t \left(\frac{2y_1'(u)}{y_1(u)}+p(u)\right)du} dv \right| ds + k \\ &= e^{mx}. \int_a^x e^{-\int_a^s \left(\frac{2me^{mu}}{e^{mu}}+b\right)du}. \left[x_0 + \int_a^s \frac{f(v)}{e^{mv}} e^{\int_a^v \left(\frac{2me^{mu}}{e^{mu}}+b\right)du} dv\right] ds + k \\ &= e^{mx}. \int_a^x e^{-(2m+b)(s-a)}. \left[x_0 + \int_a^s f(v).\ e^{-mv} e^{(2m+b)(v-a)} dv\right] ds + k \\ &= e^{mx}. \int_a^x e^{-(2m+b)(s-a)}. \left[x_0 + \int_a^s f(v).\ e^{v(m+b)-a(2m+b)} dv\right] ds + k \end{aligned}$$

for all $x \in I$ *and hence*

$$\begin{aligned} \mid y(x) - y_0(x) \mid \leq \mid e^{mx} \mid \int_a^x e^{-(2m+b)(s-a)} \\ \cdot \left| \int_s^b e^{(2m+b)(t-a)}. \frac{\varphi(t)}{\mid e^{mt} \mid} dt \right| ds. \end{aligned}$$

Example 8 *Consider* (3.24) *when* $b^2 - 4c < 0$, *with* $m = \frac{-b\pm\sqrt{b^2-4c}}{2} = \alpha \pm i\beta$. *Let* $f : I \to \mathbb{R}$, $\varphi : I \to [0, \infty)$ *be continuous function. Assume that* $y : I \to \mathbb{R}$ *is a twice continuously differential function satisfying the differential inequality*

$$\mid y'' + by' + cy - f(x) \mid \leq \varphi(x)$$

for all $x \in I$. *Since* $y_1(x) = e^{\alpha x}\cos(\beta x)$ *is a solution of* (3.24), *then using Theorem 39 that there exists a solution* $y_0 : I \to \mathbb{R}$ *of* (3.24) *such that*

$$\begin{aligned} y_0(x) = y_1(x) \cdot \int_a^x e^{-\Re \int_a^s \left(\frac{2y_1'(u)}{y_1(u)}+p(u)\right)du} \\ \cdot \left| x_0 + \int_b^s e^{\Re \int_a^t \left(\frac{2y_1'(u)}{y_1(u)}+p(u)\right)du} dv \right| ds + k \end{aligned}$$

$$= e^{\alpha x}\cos(\beta x).\int_a^x e^{-\int_a^s\left(\frac{2[\alpha\cos(\beta u)e^{\alpha u}-\beta e^{\alpha u}\sin(\beta u)]}{e^{\alpha u}\cos(\beta u)}+b\right)du}$$

$$.\left[x_0+\int_a^s \frac{f(v)}{e^{\alpha v}\cos(\beta v)}e^{\int_a^v\left(\frac{2\alpha\cos(\beta u)e^{\alpha u}-2\beta e^{\alpha u}\sin(\beta u)}{e^{\alpha u}\cos(\beta u)}+b\right)du}dv\right]ds+k$$

$$= e^{\alpha x}\cos(\beta x)\int_a^x e^{-\int_a^s[2\alpha-2\beta\tan(\beta u)+b]du}$$

$$.\left[x_0+\int_a^s \frac{f(v)}{e^{\alpha v}\cos(\beta v)}e^{\int_a^v[2\alpha-2\beta\tan(\beta u)+b]du}dv\right]ds+k$$

$$= e^{\alpha x}\cos(\beta x)\int_a^x e^{(2\alpha+b)(a-s)+[2\ln\sec(\beta u)]_a^s}$$

$$.\left[x_0+\int_a^s \frac{f(v)}{e^{\alpha v}\cos(\beta v)}e^{(2\alpha+b)(v-a)-[2\ln\sec(\beta u)]_a^v}\right]ds+k$$

$$= e^{\alpha x}\cos(\beta x)\int_a^x e^{(2\alpha+b)(a-s)+[\ln\sec^2(\beta s)-\ln\sec^2(\beta a)]}$$

$$.\left[x_0+\int_a^s \frac{f(v)}{e^{\alpha v}\cos(\beta v)}e^{(2\alpha+b)(v-a)-[\ln\sec^2(\beta v)-\ln\sec^2(\beta a)]}dv\right]ds+k$$

$$= e^{\alpha x}\cos(\beta x)\int_a^x e^{(2\alpha+b)(a-s)}\ .\ e^{\ln\frac{\sec^2(\beta s)}{\sec^2(\beta a)}}$$

$$.\left[x_0+\int_a^s \frac{f(v)}{e^{\alpha v}\cos(\beta v)}e^{(2\alpha+b)(v-a)}\ .\ e^{\ln\frac{\sec^2(\beta a)}{\sec^2(\beta v)}}dv\right]ds+k$$

$$= e^{\alpha x}\cos(\beta x)\int_a^x e^{(2\alpha+b)(a-s)}\ .\ \frac{\sec^2(\beta s)}{\sec^2(\beta a)}$$

$$.\left[x_0+\int_a^s \frac{f(v)}{\cos(\beta v)}e^{-\alpha v}e^{(2\alpha+b)(v-a)}\ .\ \frac{\sec^2(\beta a)}{\sec^2(\beta v)}dv\right]ds+k$$

$$= e^{\alpha x}\cos(\beta x)\ .\ cos^2(\beta a)\int_a^x \frac{e^{(2\alpha+b)(a-s)}}{\cos^2(\beta s)}$$

$$.\left[x_0+\int_a^s \frac{f(v)}{\cos(\beta v)}e^{-\alpha v}\ .\ \cos^2(\beta v)\ .\ \frac{e^{(2\alpha+b)(v-a)}}{\cos^2(\beta a)}\right]ds+k$$

$$= e^{\alpha x}\cos(\beta x)\ .\ x_0\cos^2(\beta a)\int_a^x \frac{e^{(2\alpha+b)(a-s)}}{\cos^2(\beta s)}ds$$

$$+e^{\alpha x}\cos(\beta x)\ .\ \int_a^s \frac{e^{-(2\alpha+b)s}}{\cos^2(\beta s)}\ .\ \int_a^s f(v)\ .\ e^{\alpha v}\cos(\beta v)e^{(2\alpha+b)v}ds$$

$$= e^{\alpha x}\cos(\beta x)\ .\ x_0\cos^2(\beta a)\int_a^x \frac{e^{(2\alpha+b)(a-s)}}{\cos^2(\beta s)}ds$$

$$+e^{\alpha x}\cos(\beta x)\ .\ \int_a^x \frac{e^{-(2\alpha+b)s}}{\cos^2(\beta s)}\ .\ \int_a^s f(v)\ .\ e^{v(\alpha+b)}\ .\ \cos(\beta v)ds,$$

where $k = \frac{y(a)}{e^{\alpha a}\cos(\beta a)}$, $x \in I$ *and* $x_0 \in \mathbb{R}$, *then*

$$| y(x) - y_0(x) | \leq | e^{\alpha x}\cos(\beta x) | \ . \int_a^x \frac{e^{(2\alpha+b)(a-s)}}{\cos^2(\beta s)}$$

$$. \left| \int_s^b \cos^2(\beta t) \ . \ e^{(\alpha+b)t} \ . \ \varphi(t)dt \right| ds.$$

3.6 Notes

The results in this chapter are adopted from the works [7], [14], [29]–[32] and [40].

Chapter 4

Hyers-Ulam Stability of Exact Linear Differential Equations

This chapter deals with the Hyers-Ulam stability of exact second order differential equations. Indeed, a second order differential equation is of the form:

$$p_0(x)y'' + p_1(x)y' + p_2(x)y + f(x) = 0 \tag{4.1}$$

is said to be exact if

$$p_0''(x) - p_1'(x) + p_2(x) = 0, \tag{4.2}$$

where $p_0(x), p_1(x)$ and $p_2(x)$ are real valued continuous functions.

4.1 Hyers-Ulam Stability of $p_0(x)y'' + p_1(x)y' + p_2(x)y + f(x) = 0$

Theorem 40 *Let $p_0, p_1, p_2, f : I \to \mathbb{R}$ be continuous function with $p_0(x) \neq 0$ for all $x \in I$, and let $\varphi : I \to [0, \infty)$ be a function. Assume that $y : I \to \mathbb{R}$ is a twice continuously differentiable function satisfying the differential inequality*

$$| p_0(x)y'' + p_1(x)y' + p_2(x)y + f(x) | \leq \varphi(x) \tag{4.3}$$

for all $x \in I$ and (4.2) is true. Then there exists a solution $y_0 : I \to \mathbb{R}$ of (4.1), such that

$$| y(x) - y_0(x) |$$
$$\leq e^{-\int_a^x p(u)du} \int_x^b \left(| p_0(v) |^{-1} \int_a^v \varphi(t)dt \right) e^{\int_a^v p(u)du} dv$$

for all $x \in I$ and $\varphi(x)e^{\int_a^x p(u)du}$ is integrable on I, where $p(x) = [p_0(x)]^{-1}[p_1(x) - p_0'(x)]$.

Proof 68 *It follows from (4.2) and (4.3) that*

$$\begin{aligned}\varphi(x) &\geq |\, p_0(x)y'' + p_1(x)y' + p_2(x)y + f(x)\,| \\ &= |\, p_0(x)y'' + p_1(x)y' + p_2(x)y + f(x) + p_0'(x)y' - p_0'(x)y' + p_0''(x)y \\ &\quad - p_0''(x)y + p_1'(x)y - p_1'(x)y\,| \\ &= |\, \{p_0(x)y'' + p_0'(x)y' - p_0''(x)y - p_0'(x)y'\} + \{p_1'(x)y + p_1(x)y'\} \\ &\quad + p_0''(x)y - p_1'(x)y + p_2(x)y + f(x)\,| \\ &= |\, [p_0(x)y' - p_0'(x)y]' + [p_1(x)y]' + [p_0''(x) - p_1'(x) + p_2(x)]y + f(x)\,| \\ &= |\, [p_0(x)y' - p_0'(x)y]' + [p_1(x)y]' + f(x)\,|,\end{aligned}$$

that is,

$$-\varphi(x) \leq [p_0(x)y' - p_0(x)y]' + [p_1(x)y]' + f(x) \leq \varphi(x). \tag{4.4}$$

Integrating (4.4) from a to x for each $x \in I$, we get

$$\begin{aligned}-\int_a^x \varphi(x)dx \leq \int_a^x [p_0(x)y' - p_0'(x)y]'dx + \int_a^x [p_1(x)y]'dx + \int_a^x f(x)dx \\ \leq \int_a^x \varphi(x)dx,\end{aligned}$$

that is,

$$-\int_a^x \varphi(t)dt \leq [p_0(x)y' - p_0'(x)y]_a^x + [p_1(x)y]_a^x + \int_a^x f(t)dt \leq \int_a^x \varphi(t)dt$$

implies that

$$\begin{aligned}-\int_a^x \varphi(t)dt \leq \{p_0(x)y' - p_0'(x)y - p_0(a)y' + p_0(a)'y\} + \{p_1(x)y - p_1(a)y\} \\ + \int_a^x f(t)dt \leq \int_a^x \varphi(t)dt,\end{aligned}$$

that is,

$$\begin{aligned}-\int_a^x \varphi(t)dt \leq \{p_0(x)y' - p_0'(x)y + p_1(x)y\} + \{p_0(a)'y - p_0(a)y' - p_1(a)y\} \\ + \int_a^x f(t)dt \leq \int_a^x \varphi(t)dt.\end{aligned}$$

Hence,

$$-\int_a^x \varphi(t)dt \leq p_0(x)y' - p_0'(x)y + p_1(x)y + k + \int_a^x f(t)dt \leq \int_a^x \varphi(t)dt.$$

Therefore,

$$\left| p_0(x)y' - p_0'(x)y + p_1(x)y + k + \int_a^x f(t)dt \right| \leq \int_a^x \varphi(t)dt$$

implies that

$$\left|p_0(x)\left\{y' + (p_0(x))^{-1}(p_1(x) - p_0'(x))y + (p_0(x))^{-1}\left(k + \int_a^x f(t)dt\right)\right\}\right| \leq \int_a^x \varphi(t)dt. \quad (4.5)$$

As a result,

$$\left|y' + (p_0(x))^{-1}[p_1(x) - p_0'(x)]y + (p_0(x))^{-1}\left(k + \int_a^x f(t)dt\right)\right| \leq | p_0(x) |^{-1} \int_a^x \varphi(t)dt. \quad (4.6)$$

Let $p(x) = (p_0(x))^{-1}[p_1(x) - p_0'(x)]$, $h(x) = (p_0(x))^{-1}\left(k + \int_a^x f(t)dt\right)$ *and* $\varphi_1(x) = | p_0(x) |^{-1} \int_a^x \varphi(t)dt$. *Then (4.6) becomes*

$$| y' + p(x)y + h(x) | \leq \varphi_1(x).$$

Let us consider,

$$z(x) = e^{\int_a^x p(u)du} y(x) + \int_a^x e^{\int_a^v p(u)du} h(v)dv$$

for all $x, t \in I$. *Hence,*

$$\begin{aligned}
| z(x) - z(t) | &= | e^{\int_a^x p(u)du} y(x) - e^{\int_a^t p(u)du} y(t) + \int_a^x e^{\int_a^v p(u)du} h(v)dv \\
&\quad - \int_a^t e^{\int_a^v p(u)du} h(v)dv | \\
&= \left| e^{\int_a^x p(u)du} y(x) - e^{\int_a^s p(u)du} y(t) + \int_t^x e^{\int_a^v p(u)du} h(v)dv \right| \\
&= \left| \int_t^x \frac{d}{dv}[e^{\int_a^v p(u)du} y(v)]dv + \int_t^x e^{\int_a^v p(u)du} h(v)dv \right| \\
&= \left| \int_t^x e^{\int_a^v p(u)du} y(v)p(v) + y'(v)e^{\int_a^v p(u)du} + \int_t^x e^{\int_a^v p(u)du} h(v)dv \right| \\
&= \left| \int_t^x e^{\int_a^v p(u)du} [y'(v) + p(v)y(v) + h(v)]dv \right| \\
&\leq \left| \int_t^x e^{\int_a^v p(u)du} \varphi_1(v)dv \right|.
\end{aligned}$$

Since $\varphi(x)e^{\int_a^x p(u)du}$ *is integrable on* I, *then we can find* $x_0 \in I$ *for* $\epsilon > 0$ *such that* $| z(x) - z(t) | < \epsilon$ *for* $x, t \geq x_0$, *that is,* $z(t)_{t \in I}$ *is a Cauchy net and hence there exists a* $z \in \mathbb{R}$ *such that* $z(t)$ *converges to* z *as* $t \to b$.

$$\begin{aligned}
&\left|y(x) - e^{-\int_a^x p(u)du}\left[z - \int_a^x e^{\int_a^v p(u)du}h(v)dv\right]\right| \\
&= \left|e^{-\int_a^x p(u)du}[z(x) - z]\right| \\
&\leq e^{-\int_a^x p(u)du}|z(x) - z(t)| + e^{-\int_a^x p(u)du}|z(t) - z| \\
&\leq e^{-\int_a^x p(u)du}\left|\int_t^x e^{\int_a^v p(u)du}\varphi_1(v)dv\right| + e^{-\int_a^x p(u)du}|z(t) - z| \\
&\leq e^{-\int_a^x p(u)du}\left|\int_b^x e^{\int_a^v p(u)du}\varphi_1(v)dv\right|,
\end{aligned}$$

where $z(t) \to z$ *as* $t \to b$. *If*

$$y_0(x) = e^{-\int_a^x p(u)du}\left[z - \int_a^x e^{\int_a^v p(u)du}h(v)dv\right],$$

then the last inequality becomes

$$\begin{aligned}
&|y(x) - y_0(x)| \\
&\leq \left|e^{-\int_a^x p(u)du}\int_x^b \varphi_1(v)e^{\int_a^v p(u)du}dv\right| \\
&\leq \left|e^{-\int_a^x p(u)du}\int_x^b \left(|\, p_0(v)\,|^{-1}\int_a^v \varphi(t)dt\right)e^{\int_a^v p(u)du}dv\right|
\end{aligned}$$

for all $x \in I$.

Corollary 23 *Let* p_0, p_1, p_2, $\mu : I \to \mathbb{R}$ *be continuous functions with* $p_0(x) \neq 0$ *and* $\mu(x) \neq 0$ *for all* $x \in I$, *and let* $\varphi : I \to [0,\infty)$ *be a function. Assume that* $y : I \to \mathbb{R}$ *is a twice continuously differentiable function satisfying the differential inequality*

$$|\, \mu(x)[p_0(x)y'' + p_1(x)y' + p_2(x)y + f(x)] \,| \leq \varphi(x) \tag{4.7}$$

for all $x \in I$ *and* $\{\mu(x)p_0(x)\}'' - \{p_1(x)\mu(x)\}' + p_2(x)\mu(x) = 0$ *is true. Then there exists a solution* $y_0 : I \to \mathbb{R}$ *of* $\mu(x)[p_0(x)y''+p_1(x)y'+p_2(x)y+f(x)] = 0$ *such that*

$$\begin{aligned}
&|y(x) - y_0(x)| \\
&\leq e^{-\int_a^x p(u)du}\int_x^b \left(|\mu(v)p_0(v)|^{-1}\int_a^v \varphi(t)dt\right)e^{\int_a^v p(u)du}dv
\end{aligned}$$

for $x \in I$, *where* $p(x) = \{\mu(x)p_0(x)\}^{-1}[\mu(x)p_1(x) - \{\mu(x)p_0(x)\}']$.

Proof 69 *Indeed, the equation*

$$\mu(x)[p_0(x)y'' - p_1(x)y' + p_2(x)y + f(x)] = 0$$

is said to be exact if

$$\{\mu(x)p_0(x)\}'' - \{p_1(x)\mu(x)\}' + p_2(x)\mu(x) = 0. \tag{4.8}$$

It follows from (4.7) and (4.8) that

$$\begin{aligned}
\varphi(x) &\geq |\ \mu(x)p_0(x)y'' + \mu(x)p_1(x)y' + \mu(x)p_2(x)y + \mu(x)f(x)\ | \\
&= |\ \{\mu(x)p_0(x)\}y'' + \{\mu(x)p_1(x)\}y' + \{\mu(x)p_2(x)\}y + \ \mu(x)f(x) \\
&\quad \{\mu(x)p_0(x)\}'y' - \{\mu(x)p_0(x)\}'y' + \{\mu(x)p_0(x)\}''y - \{\mu(x)p_0(x)\}''y \\
&\quad + \{\mu(x)p_1(x)\}'y - \{\mu(x)p_1(x)\}'y\ | \\
&= |\ [\{\mu(x)p_0(x)\}y'' + \{\mu(x)p_0(x)\}'y' - \{\mu(x)p_0(x)\}''y \\
&\quad - \{\mu(x)p_0(x)\}'y + [\{\mu(x)p_1(x)\}'y + \{\mu(x)p_1(x)\}y'] \\
&\quad + [\{\mu(x)p_0(x)\}''y - \{\mu(x)p_1(x)\}'y + \{\mu(x)p_2(x)\}y] + \mu(x)f(x)\ | \\
&= |\ [\{\mu(x)p_0(x)\}y' - \{\mu(x)p_0(x)\}'y]' + [\{\mu(x)p_1(x)\}y]' \\
&\quad + [\{\mu(x)p_0(x)\}'' - \{\mu(x)p_1(x)\}' + \{\mu(x)p_2(x)\}]y + \mu(x)f(x)\ | \\
&= |\ [\{\mu(x)p_0(x)\}y' - \{\mu(x)p_0(x)\}'y]' + [\{\mu(x)p_1(x)\}y]' + \mu(x)f(x)\ |,
\end{aligned}$$

that is,

$$\begin{aligned}
-\varphi(x) \leq [\{\mu(x)p_0(x)\}y' - \{\mu(x)p_0(x)\}'y]' + [\{\mu(x)p_1(x)\}y]' \\
+ \mu(x)f(x) \leq \varphi(x).
\end{aligned} \tag{4.9}$$

Integrating (4.9) from a to x for each $x \in I$, we get

$$\begin{aligned}
-\int_a^x \varphi(x)dx \leq \int_a^x [\{\mu(x)p_0(x)\}y' - \{\mu(x)p_0(x)\}'y]'dx + \int_a^x \{\mu(x)p_1(x)\}y]'dx \\
+ \int_a^x \mu(x)f(x)dx \leq \int_a^x \varphi(x)dx,
\end{aligned}$$

that is,

$$\begin{aligned}
-\int_a^x \varphi(t)dt \leq [\{\mu(x)p_0(x)\}y' - \{\mu(x)p_0(x)\}'y]_a^x + [\{\mu(x)p_1(x)\}y]_a^x \\
+ \int_a^x \mu(t)f(t)dt \leq \int_a^x \varphi(t)dt
\end{aligned}$$

implies that

$$\begin{aligned}
-\int_a^x \varphi(t)dt \leq \{\mu(x)p_0(x)\}y' - \{\mu(x)p_0(x)\}'y - \{\mu(a)p_0(a)\}y' + \{\mu(a)p_0(a)\}'y \\
+ \{\mu(x)p_1(x)\}y - \{\mu(a)p_1(a)\}y + \int_a^x \mu(t)f(t)dt \leq \int_a^x \varphi(t)dt.
\end{aligned}$$

Therefore,

$$-\int_a^x \varphi(t)dt \le \{\mu(x)p_0(x)\}y' - \{\mu(x)p_0(x)\}'y + \{\mu(x)p_1(x)y\} + \{\mu(a)p_0(a)\}'y$$
$$- \{\mu(a)p_0(a)\}y' - \{\mu(a)p_1(a)\}y\} + \int_a^x \mu(t)f(t)dt \le \int_a^x \varphi(t)dt.$$

Hence,

$$\left|\{\mu(x)p_0(x)\}y' - \{\mu(x)p_0(x)\}'y + \{\mu(x)p_1(x)\}y + k + \int_a^x \mu(t)f(t)dt\right|$$
$$\le \int_a^x \varphi(t)dt,$$

that is,

$$\left|\{\mu(x)p_0(x)\}\left\{y' + \{\mu(x)p_0(x)\}^{-1}[\mu(x)p_1(x) - \{\mu(x)p_0(x)\}']y\right.\right.$$
$$\left.\left. + \{\mu(x)p_0(x)\}^{-1}\left(k + \int_a^x \mu(t)f(t)dt\right)\right\}\right| \le \int_a^x \varphi(t)dt$$

and hence

$$\left|y' + \{\mu(x)p_0(x)\}^{-1}[\mu(x)p_1(x) - \{\mu(x)p_0(x)\}']y + \{\mu(x)p_0(x)\}^{-1}\right.$$
$$\left.\left(k + \int_a^x \mu(t)f(t)dt\right)\right| \le |\mu(x)p_0(x)|^{-1}\int_a^x \varphi(t)dt. \quad (4.10)$$

If we choose, $p(x) = \{\mu(x)p_0(x)\}^{-1}[\mu(x)p_1(x) - \{\mu(x)p_0(x)\}']$, $\varphi_1(x) = |\,\mu(x)p_0(x)\,|^{-1}\int_a^x \varphi(t)dt$ *and* $h(x) = \{\mu(x)p_0(x)\}^{-1}\left(k + \int_a^x \mu(t)f(t)dt\right)$, *where* $k = -[\{\mu(a)p_0(a)\}y'(a) - \{\mu(a)p_0(a)\}'y(a) + \{\mu(a)p_1(a)\}y(a)]$, *then (4.10) becomes*

$$|\, y' + p(x)y + h(x)\,| \le \varphi_1(x).$$

By using Theorem 40, there exists a unique $z \in \mathbb{R}$ *such that*

$$y_0(x) = e^{-\int_a^x p(u)du}\left[z - \int_a^x e^{\int_a^v p(u)du}h(v)dv\right].$$

Thus,

$$|\, y(x) - y_0(x)\,|$$
$$\le e^{-\int_a^x p(u)du}\int_x^b \left(|\,\mu(v)p_0(v)\,|^{-1}\int_a^v \varphi(t)dt\right)e^{\int_a^v p(u)du}dv.$$

Corollary 24 *Consider the equation* $y'' + cy' + by + f(x) = 0$. *Let* $c^2 - 4b \ge 0$, $m = \frac{c\pm\sqrt{c^2-4b}}{2}$, $f : I \to \mathbb{R}$ *be a continuous function and let* $\varphi : I \to [0, \infty)$

be a function. Assume that $y : I \to \mathbb{R}$ is a twice continuously differentiable function satisfying the differential inequality

$$| y'' + cy' + by + f(x) | \leq \varphi(x) \tag{4.11}$$

for all $x \in I$. Then there exists a solution $y_0 : I \to \mathbb{R}$ of $y'' + cy' + by + f(x) = 0$ such that

$$| y(x) - y_0(x) | \leq e^{(m-c)(x-a)} \int_x^b \left(e^{(-mv)} \int_a^v e^{(mt)} \varphi(t) \right) e^{(c-m)(x-a)} dv$$

for all $x \in I$.

Proof 70 *For the second order linear differential equation*

$$y'' + cy' + by + f(x) = 0, \tag{4.12}$$

if we choose $\mu(x)$ is an integrating factor of (4.12), then it follows that

$$\mu''(x) - c\mu'(x) + b\mu(x) = 0.$$

Consider $\mu(x) = e^{mx}$ is an integrating factor of (4.12) when $c^2 - 4b \geq 0$ with $m = \frac{c \pm \sqrt{c^2-4b}}{2}$, then (4.11) becomes

$$e^{mx} | y'' + cy' + by + f(x) | \leq e^{mx} \varphi(x)$$

for all $x \in I$. Using Corollary 23 with $\varphi_1(x) = e^{mx}\varphi(x)$,

$$h(x) = \{\mu(x)p_0(x)\}^{-1} \left(k + \int_a^x \mu(t) f(t) dt \right) = (e^{mx})^{-1} \left(k + \int_a^x e^{mt} f(t) dt \right)$$

and

$$\begin{aligned} p(x) &= \{\mu(x)p_0(x)\}^{-1} [\mu(x)p_1(x) - \{\mu(x)p_0(x)\}'] \\ &= (e^{mx})^{-1} [e^{mx} c - (e^{mx} \ . \ 1)'] \\ &= \frac{1}{e^{mx}} (e^{mx} c - m e^{mx}) \\ &= c - m, \end{aligned}$$

there exists a unique $z \in \mathbb{R}$ such that

$$\begin{aligned} y_0(x) &= e^{-\int_a^x p(u)du} \left[z - \int_a^x e^{\int_a^v p(u)du} h(v) dv \right] \\ &= e^{-\int_a^x (c-m)du} \left[z - \int_a^x e^{\int_a^v (c-m)du} \{e^{mv}\}^{-1} \left(k + \int_a^v e^{mx} f(x) dx \right) dv \right] \\ &= e^{(m-c)(x-a)} \left[z - \int_a^x e^{(c-m)(v-a)} . e^{-mv} \left(k + \int_a^v e^{mx} f(x) dx \right) dv \right] \\ &= e^{(m-c)(x-a)} \left[z - \int_a^x e^{v(c-2m) - a(c-m)} \left(k + \int_a^v e^{mx} f(x) dx \right) dv \right], \end{aligned}$$

where $k = -[e^{ma}y'(a) - me^{ma}y(a) + ce^{ma}y(a)]$ *for all* $x \in I$ *and hence*

$$\begin{aligned}
\mid y(x) - y_0(x) \mid &\leq e^{-\int_a^x p(u)du} \int_x^b \left(\mid \mu(v)p_0(v) \mid^{-1} \int_a^v \varphi(t)dt \right) e^{\int_a^v p(u)du} dv \\
&\leq e^{-\int_a^x (c-m)du} \int_x^b \left(\mid e^{mv} \mid^{-1} \int_a^v \varphi(t)dt \right) e^{\int_a^v (c-m)du} dv \\
&\leq e^{(m-c)(x-a)} \int_x^b \left(e^{mv} \int_a^v e^{mt}\varphi(t) \right) e^{(c-m)(v-a)} dv.
\end{aligned}$$

Corollary 25 *Consider* (4.12). *Let* $c^2 - 4b < 0$, $m = \frac{c \pm \sqrt{c^2-4b}}{2} = \alpha \pm i\beta$, $f : I \to \mathbb{R}$ *be a continuous function and let* $\varphi : I \to [0, \infty)$ *be a function. Assume that* $y : I \to \mathbb{R}$ *is a twice continuously differentiable function satisfying the differential inequality* (4.11) *for all* $x \in I$. *Then there exists a solution* $y_0 : I \to \mathbb{R}$ *of* (4.12) *such that*

$$\begin{aligned}
&\mid y(x) - y_0(x) \mid \\
&\leq e^{-\int_a^x p(u)du} \int_x^b \left(\mid \mu(v) \mid^{-1} \int_a^v e^{\alpha t}\varphi(t)dt \right) e^{\int_a^v p(u)du} dv
\end{aligned}$$

for all $x \in I$, *where* $\mu(x) = e^{\alpha x} \cos \beta x$ *and* $p(x) = [c - \alpha + \beta \tan \beta x]$.

Proof 71 *Consider* (4.12) *when* $c^2 - 4b < 0$ *with* $m = \dfrac{c \pm \sqrt{c^2 - 4b}}{2} = \alpha \pm i\beta$, $\mu(x) = e^{\alpha x} \cos \beta x$ *is an integrating factor. Then the inequality* (4.11) *becomes*

$$e^{\alpha x} \mid \cos \beta x \{y'' + cy' + by + f(x)\} \mid \leq e^{\alpha x} \mid y'' + cy' + by + f(x) \mid \leq e^{\alpha x}\varphi(x)$$

for all $x \in I$. *By Corollary 24 with*

$$\begin{aligned}
h(x) &= \{\mu(x)p_0(x)\}^{-1} \left(k + \int_a^x \mu(t)f(t)dt \right) \\
&= (e^{\alpha x} \cos \beta x)^{-1} \left(k + \int_a^x e^{\alpha t} \cos \beta t f(t)dt \right),
\end{aligned}$$

and

$$\begin{aligned}
p(x) &= \{\mu(x).1\}^{-1}[\mu(x)c - \{\mu(x).1\}'] \\
&= (e^{\alpha x} \cos \beta x)^{-1}[e^{\alpha x} \cos \beta x.c - (e^{\alpha x} \cos \beta x)'] \\
&= \frac{1}{e^{\alpha x} \cos \beta x}[e^{\alpha x} \cos \beta x.c - \alpha \cos \beta x e^{\alpha x} + \beta e^{\alpha x} \sin \beta x] \\
&= c - \alpha + \beta \tan \beta x,
\end{aligned}$$

there exists a unique $z \in \mathbb{R}$ *such that*

$$\begin{aligned}
y_0(x) &= e^{-\int_a^x p(u)du} \left[z - \int_a^x e^{\int_a^v p(u)du} h(v)dv \right] \\
&= e^{-\int_a^x p(u)du} \left[z - \int_a^x e^{\int_a^v p(u)du} (e^{\alpha x} \cos \beta x)^{-1} \left(k + \int_a^v e^{\alpha x} \cos \beta x f(x)dx \right) \right],
\end{aligned}$$

where $k = [e^{\alpha a}\cos\beta a y'(a) - (e^{\alpha a}\cos\beta a)'y(a) + ce^{\alpha a}\cos\beta a y(a)]$ *for all* $x \in I$ *and hence*

$$\begin{aligned}|\, y(x) - y_0(x)\,| &\le e^{-\int_a^x p(u)du}\int_x^b \left(|\,\mu(v)\,|^{-1}\int_a^v \varphi(t)dt\right)e^{\int_a^v p(u)du}dv \\ &\le e^{-\int_a^x p(u)du}\int_x^b \left(|\,\mu(v)\,|^{-1}\int_a^v e^{\alpha t}\varphi(t)dt\right)e^{\int_a^v p(u)du}dv.\end{aligned}$$

Example 9 *Consider the*

$$x^2y'' + \alpha xy' + \beta y + f(x) = 0,$$

where α *and* β *are real constants. It is exact when* $\alpha - \beta = 2$*. By using Theorem 40, it has the Hyers-Ulam stability.*

Let us consider $\mu(x)$ *be an integrating factor of the Euler differential equation. It is said to be exact if*

$$\{x^2\mu(x)\}'' - \{\alpha x\mu(x)\}' + \beta\mu(x) = 0,$$

that is,

$$2\mu(x) + 2x\mu'(x) + x^2\mu''(x) + 2x\mu'(x) - \alpha x\mu'(x) - \alpha\mu(x) + \beta\mu(x) = 0$$

implies that

$$x^2\mu''(x) + (4-\alpha)x\mu'(x) + (2-\alpha+\beta)\mu(x) = 0. \tag{4.13}$$

Let $\mu(x) = x^m$, $\mu'(x) = mx^{m-1}$ *and* $\mu''(x) = m(m-1)x^{m-2}$*. Then* (4.13) *becomes*

$$x^2(m^2-m)x^{m-2} + (4-\alpha)xmx^{m-1} + (2-\alpha+\beta)x^m = 0,$$

that is,

$$m^2 + (3-\alpha)m + (2-\alpha+\beta) = 0. \tag{4.14}$$

From (4.14), we obtain that

$$\begin{aligned}m &= \frac{-(3-\alpha) \pm \sqrt{(3-\alpha)^2 - 4(2-\alpha+\beta)}}{2} \\ &= \frac{-(3-\alpha) \pm \sqrt{(1-\alpha)^2 - 4\beta}}{2}.\end{aligned}$$

When $(1-\alpha)^2 - 4\beta \ge 0$*, by Corollary 24 with*

$$\begin{aligned}p(x) &= \{\mu(x)p_0(x)\}^{-1}[\mu(x)p_1(x) - \{\mu(x)p_0(x)\}'] \\ &= (e^{mx}.x^2)^{-1}[e^{mx}.\alpha x - (e^{mx}.x^2)'] \\ &= \frac{1}{e^{mx}.x^2}[e^{mx}.\alpha x - 2xe^{mx} - me^{mx}.x^2] \\ &= \frac{\alpha}{x} - \frac{2}{x} - m\end{aligned}$$

and

$$h(x) = \{\mu(x)p_0(x)\}^{-1}\left(k + \int_a^x \mu(t)f(t)dt\right)$$
$$= (e^{mx}.x^2)^{-1}\left(k + \int_a^x e^{mt}f(t)dt\right),$$

there exists a unique $z \in \mathbb{R}$ *such that*

$$y_0(x) = e^{-\int_a^x p(u)du}\left[z - \int_a^x e^{\int_a^v p(u)du}h(v)dv\right]$$
$$= e^{-\int_a^x p(u)du}\left[z - \int_a^x e^{\int_a^v p(u)du}(e^{mv}.v^2)^{-1}\left(k + \int_a^v e^{mx}f(x)dx\right)\right]$$

for all $x \in I$ *and hence*

$$| y(x) - y_0(x) | \leq e^{-\int_a^x p(u)du}\int_x^b \left(| \mu(v)p_0(v) |^{-1}\int_a^v \varphi(t)dt\right)e^{\int_a^v p(u)du}dv$$
$$\leq e^{-\int_a^x p(u)du}\int_x^b \left(| \mu(v)p_0(v) |^{-1}\int_a^v e^{mt}\varphi(t)\right)e^{\int_a^v p(u)du}dv.$$

When $(1-\alpha)^2 - 4\beta < 0$, *by Corollary 25,*

$$p(x) = \{\mu(x)p_0(x)\}^{-1}[\mu(x)p_1(x) - \{\mu(x)p_0(x)\}']$$
$$= (e^{\alpha x}\cos\beta x.x^2)^{-1}[e^{\alpha x}\cos\beta x.\alpha x - ((e^{\alpha x}\cos\beta x.x^2)']$$
$$= \frac{1}{e^{\alpha x}\cos\beta x.x^2}[e^{\alpha x}\cos\beta x.\alpha x - e^{\alpha x}\cos\beta x.2x$$
$$- x^2(\alpha e^{\alpha x}\cos\beta x - \beta e^{\alpha x}\sin\beta x)]$$
$$= \frac{1}{e^{\alpha x}\cos\beta x.x^2}[e^{\alpha x}\cos\beta x.\alpha x - 2xe^{\alpha x}\cos\beta x$$
$$- \alpha x^2 e^{\alpha x}\cos\beta x + \beta x^2 e^{\alpha x}\sin\beta x]$$
$$= \frac{\alpha}{x} - \frac{2}{x} - \alpha + \beta\tan\beta x,$$

and

$$h(x) = \{\mu(x)p_0(x)\}^{-1}\left(k + \int_a^x \mu(t)f(t)dt\right)$$
$$= (e^{\alpha x}\cos\beta x.x^2)^{-1}\left(k + \int_a^x e^{\alpha t}\cos\beta t f(t)dt\right),$$

there exists a unique $z \in \mathbb{R}$ *such that*

$$y_0(x) = e^{-\int_a^x p(u)du}\left[z - \int_a^x e^{\int_a^v p(u)du}h(v)dv\right]$$
$$= e^{-\int_a^x p(u)du}\Bigg[z - \int_a^x e^{\int_a^v p(u)du}(e^{mv}\cos\beta v.v^2)^{-1}$$
$$\left(k + \int_a^v e^{mx}\cos\beta x f(x)dx\right)\Bigg]$$

for all $x \in I$ *and hence*

$$| y(x) - y_0(x) | \leq e^{-\int_a^x p(u)du} \int_x^b \left(| \mu(v)p_0(v) |^{-1} \int_a^v \varphi(t)dt \right) e^{\int_a^v p(u)du} dv$$

$$\leq e^{-\int_a^x p(u)du} \int_x^b \left(| \mu(v)p_0(v) |^{-1} \int_a^v e^{\alpha t}\varphi(t)dt \right) e^{\int_a^v p(u)du} dv.$$

□

4.2 Notes

The results in this chapter are adopted from the work [9].

Chapter 5

Hyers-Ulam Stability of Euler's Differential Equations

This chapter deals with the Hyers-Ulam stability of Euler's differential equations of the following forms:

$$
\begin{aligned}
&ty'(t) + \alpha y(t) + \beta t^r x_0 = 0, \\
&t^2 y''(t) + \alpha t y'(t) + \beta y(t) = 0, \\
&t^3 y''' + \alpha t^2 y'' + \beta t y' + \gamma y = 0 \\
&t^4 y^{(iv)} + \alpha t^3 y''' + \beta t^2 y'' + \gamma t y' + \delta y = 0,
\end{aligned}
$$

where $t \in I = (a, b)$, $-\infty < a < b < +\infty$ and $\alpha, \beta, \gamma, \delta \in \mathbb{C}$.

5.1 Hyers-Ulam Stability of $ty'(t) + \alpha y(t) + \beta t^r x_0 = 0$

Consider

$$ty'(t) + \alpha y(t) + \beta t^r x_0 = 0. \tag{5.1}$$

Let $I = (a, b)$ be an open interval with either $0 < a < b \leq \infty$ or $\infty < a < b < 0$. If $x_0 = 1$ in the equation (5.1), then the function

$$
y(t) = \begin{cases} \dfrac{c}{t^\alpha} - \dfrac{\beta}{\alpha + r} t^r & (\text{for } \alpha + r \neq 0), \\ \dfrac{c - \beta \ln | t |}{t^\alpha} & (\text{for } \alpha + r = 0), \end{cases}
$$

where c is an arbitrary real number, is the general solution of (5.1).

Theorem 41 *Let X be a complex Banach space and let $I = (a, b)$ be an open interval as above. Assume that a function $\varphi : I \to [0, \infty)$ is given, α, β, r are complex constants,and that x_0 is a fixed element of X. Furthermore, suppose a continuously differentiable function $f : I \to X$satisfies the differential inequality*

$$|ty'(t) + \alpha y(t) + \beta t^r x_0| \leq \varphi(t) \tag{5.2}$$

for all $t \in I$. If both $t^{\alpha+r-1}$ and $t^{\alpha-1}\varphi(t)$ are integrable on (a, c) for any c with $a < c \leq b$, then there exists a unique solution $f_0 : I \to X$ of the differential equation (5.1) such that

$$\|f(t) - f_0(t)\| \leq | t^{-\alpha} | \left| \int_t^b v^{\alpha-1}\varphi(v)dv \right| \tag{5.3}$$

for any $t \in I$.

Proof 72 *(a) For case $\alpha + r \neq 0$. Let X^* be the dual space of X, i.e., the set of all continuous linear functionals $\lambda : X \to C$. For $\lambda \in X^*$, $f_\lambda : I \to C$ defined by $f_\lambda(t) = \lambda f(t)$ and $(f_\lambda)'(t) = \lambda\{f'(t)\}$. Therefore*

$$\begin{aligned}
|t(f_\lambda)'(t) + \alpha f_\lambda(t) + \lambda(\beta t^r x_0)| &= |\lambda(tf'(t) + \alpha f(t) + \beta t^r x_0)| \\
&\leq \|\lambda\| \|tf'(t) + \alpha f(t) + \beta t^r x_0\| \\
&\leq \|\lambda\| \varphi(t),
\end{aligned}$$

for all $t \in I$. Consider

$$z(t) = \left(\frac{t}{a}\right)^\alpha f(t) + \frac{\beta}{(\alpha + r)a^\alpha}(t^{\alpha+r} - a^{\alpha+r})x_0$$

for all $s, t \in I$. Now,

$$\begin{aligned}
|\lambda\{z(t) - z(s)\}| &= \left| \lambda \left\{ \left(\frac{t}{a}\right)^\alpha f(t) + \frac{\beta}{(\alpha + r)a^\alpha}(t^{\alpha+r} - a^{\alpha+r})x_0 \right\} \right. \\
&\quad \left. - \lambda \left\{ \left(\frac{s}{a}\right)^\alpha f(s) + \frac{\beta}{(\alpha + r)a^\alpha}(s^{\alpha+r} - a^{\alpha+r})x_0 \right\} \right| \\
&= \left| \left(\frac{t}{a}\right)^\alpha f_\lambda(t) - \left(\frac{s}{a}\right)^\alpha f_\lambda(s) + \frac{\beta}{(\alpha + r)a^\alpha}(t^{\alpha+r} - s^{\alpha+r})\lambda x_0 \right| \\
&= \left| \int_s^t \frac{d}{dv}\left[\left(\frac{v}{a}\right)^\alpha f_\lambda(v)\right] dv + \int_s^t \lambda \left(\frac{\beta}{a^\alpha} v^{\alpha+r-1} x_0\right) dv \right| \\
&= \left| \int_s^t \left\{ \left(\frac{v}{a}\right)^\alpha f'_\lambda(v) + \alpha f_\lambda(v) \frac{v^{\alpha-1}}{a^\alpha} \right\} dv \right. \\
&\quad \left. + \int_s^t \lambda \left(\frac{\beta}{a^\alpha} v^{\alpha+r-1} x_0\right) dv \right| \\
&= \left| \int_s^t \left(\frac{v}{a}\right)^\alpha \left\{ f'_\lambda(v) + \frac{\alpha}{v} f_\lambda(v) + \lambda(\beta v^{r-1} x_0) \right\} dv \right| \\
&= \left| \int_s^t \frac{v^{\alpha-1}}{a^\alpha} \{v(f_\lambda)'(v) + \alpha f_\lambda(v) + \lambda(\beta v^r x_0)\} dv \right| \\
&\leq \|\lambda\| \left| \int_s^t \frac{v^{\alpha-1}}{a^\alpha} \varphi(v) dv \right|.
\end{aligned} \tag{5.4}$$

Since $\lambda \in X^*$ *is selected arbitrarily, then the inequality* (5.4) *becomes*

$$\|z(t) - z(s)\| \leq \left| \int_s^t \frac{v^{\alpha-1}}{a^\alpha} \varphi(v) dv \right|,$$

for all $s, t \in I$. *Since* $t^{\alpha-1}\varphi(t)$ *is integrable on* (a, c) *for any* c *with* $a < c \leq b$, *then we can find* $t_0 \in I$ *such that* $\|z(t) - z(s)\| < \epsilon$ *for* $s, t \geq t_0$, *that is,* $z(s)_{s \in I}$ *is a Cauchy net and hence there exists an* $x \in X$ *such that* $z(s)$ *converges to* x *as* $s \to b$, *since* X *is complete. Indeed,*

$$\begin{aligned}
&\left\| f(t) - \left(\frac{a}{t}\right)^\alpha x + \frac{\beta}{(\alpha + r)t^r}(t^{\alpha+r} - a^{a+r})x_0 \right\| \\
&= \|a^\alpha t^{-\alpha}\{z(t) - x\}\| \\
&\leq |a^\alpha t^{-\alpha}| \|z(t) - z(s)\| + |a^\alpha t^{-\alpha}| \|z(s) - x\| \\
&\leq |a^\alpha t^{-\alpha}| \left| \int_s^t \frac{v^{\alpha-1}}{a^\alpha} \varphi(v) dv \right| + |a^\alpha t^{-\alpha}| \|z(s) - x\| \\
&= |a^\alpha t^{-\alpha}| \left| \int_b^t \frac{v^{\alpha-1}}{a^\alpha} \varphi(v) dv \right|
\end{aligned}$$

as $s \to b$, $z(s) \to x$ *and thus*

$$\begin{aligned}
&\left\| f(t) - \left(\frac{a}{t}\right)^\alpha x + \frac{\beta}{(\alpha + r)t^r}(t^{\alpha+r} - a^{a+r})x_0 \right\| \\
&\qquad \leq |t^{-\alpha}| \left| \int_b^t v^{\alpha-1} \varphi(v) dv \right|. \qquad (5.5)
\end{aligned}$$

If,

$$f_0(t) = \left(\frac{a}{t}\right)^\alpha x - \frac{\beta}{(\alpha + r)t^r}(t^{\alpha+r} - a^{a+r})x_0, \qquad (5.6)$$

then (5.5) *becomes,*

$$\|f(t) - f_0(t)\| \leq |t^{-\alpha}| \left| \int_b^t v^{\alpha-1} \varphi(v) dv \right|.$$

Now, it remains to show the uniqueness of f_0. *Assume that* $x_1 \in X$ *and there exists another solution*

$$f_1(t) = \left(\frac{a}{t}\right)^\alpha x_1 - \frac{\beta}{(\alpha + r)t^r}(t^{\alpha+r} - a^{\alpha+r})x_0, \qquad t \in I.$$

So,

$$\begin{aligned}
\left\| \left(\frac{a}{t}\right)^\alpha (x_1 - x) \right\| = &\left\| f(t) - \left(\frac{a}{t}\right)^\alpha x + \frac{\beta}{(\alpha + r)t^r}(t^{\alpha+r} - a^{\alpha+r})x_0 \right. \\
&\left. - f(t) + \left(\frac{a}{t}\right)^\alpha x_1 - \frac{\beta}{(\alpha + r)t^r}(t^{\alpha+r} - a^{\alpha+r})x_0 \right\|
\end{aligned}$$

$$\leq \left\| f(t) - \left(\frac{a}{t}\right)^{\alpha} x + \frac{\beta}{(\alpha+r)t^{r}}(t^{\alpha+r} - a^{\alpha+r})x_0 \right\|$$
$$+ \left\| -\left\{ f(t) - \left(\frac{a}{t}\right)^{\alpha} x_1 + \frac{\beta}{(\alpha+r)t^{r}}(t^{\alpha+r} - a^{a+r})x_0 \right\} \right\|$$
$$\leq |t^{-\alpha}| \left| \int_b^t v^{\alpha-1}\varphi(v)dv \right| + |t^{-\alpha}| \left| \int_b^t v^{\alpha-1}\varphi(v)dv \right|$$
$$\leq 2|t^{-\alpha}| \left| \int_b^t v^{\alpha-1}\varphi(v)dv \right|$$

implies that

$$\|x_1 - x\| \leq \frac{2}{|a^{\alpha}|} \left| \int_t^b v^{\alpha-1}\varphi(v)dv \right| \to 0,$$

as $t \to b$. *Hence* $x_1 = x$.
(*b*) *For the case* $\alpha + r = 0$. *Consider*

$$z(t) = \left(\frac{t}{a}\right)^{\alpha} f(t) + \frac{\beta}{a^{\alpha}}(\ln |\, t \,| - \ln |\, a \,|)x_0$$

for all $t \in I$. *Now,*

$$|\lambda\{z(t) - z(s)\}| = \left| \lambda \left\{ \left(\frac{t}{a}\right)^{\alpha} f(t) + \frac{\beta}{a^{\alpha}}(\ln|t| - \ln|a|)x_0 \right.\right.$$
$$\left.\left. - \left(\frac{s}{a}\right)^{\alpha} f(s) - \frac{\beta}{a^{\alpha}}(\ln|s| - \ln|a|)x_0 \right\} \right|$$
$$= \left| \left(\frac{t}{a}\right)^{\alpha} f_{\lambda}(t) - \left(\frac{s}{a}\right)^{\alpha} f_{\lambda}(s) + \frac{\beta}{a^{\alpha}}(\ln|t| - \ln|s|)\lambda x_0 \right|$$
$$= \left| \int_s^t \frac{d}{dv}\left[\left(\frac{v}{a}\right)^{\alpha} f_{\lambda}(v) \right] dv + \int_s^t \lambda \frac{\beta}{a^{\alpha}} \left(\frac{1}{v}\right) x_0 dv \right|$$

$$= \left| \int_s^t \left\{ \left(\frac{v}{a}\right)^{\alpha} f'_{\lambda}(v) + \alpha f_{\lambda}(v) \frac{v^{\alpha-1}}{a^{\alpha}} \right\} dv \right.$$
$$\left. + \int_s^t \lambda \frac{\beta}{a^{\alpha}} \left(\frac{1}{v}\right) x_0 dv \right|$$
$$= \left| \int_s^t \left\{ \left(\frac{v}{a}\right)^{\alpha} v f'_{\lambda}(v) + \alpha f_{\lambda}(v) \left(\frac{v}{a}\right)^{\alpha} + \lambda \frac{\beta}{a^{\alpha}} x_0 \right\} dv \right|$$
$$= \left| \int_s^t \left(\frac{v}{a}\right)^{\alpha} \{ v f'_{\lambda}(v) + \alpha f_{\lambda}(v) + \lambda(\beta v^{-\alpha} x_0)\} dv \right|$$
$$= \left| \int_s^t \left(\frac{v}{a}\right)^{\alpha} \{ v f'_{\lambda}(v) + \alpha f_{\lambda}(v) + \lambda(\beta v^{r} x_0)\} dv \right|$$
$$\leq \|\lambda\| \left| \int_s^t \left(\frac{v}{a}\right)^{\alpha} \varphi(v) dv \right|. \tag{5.7}$$

Since $\lambda \in X^*$ is selected arbitrarily, then the inequality (5.7) becomes

$$\|z(t) - z(s)\| \leq \left| \int_s^t \frac{v^{\alpha-1}}{a^\alpha} \varphi(v) dv \right|,$$

for $s, t \in I$. Proceeding as in case (a), we choose

$$f_0(t) = \left(\frac{a}{t}\right)^\alpha x - \frac{\beta}{t^\alpha}(\ln | t | - \ln | a |)x_0.$$

Therefore,

$$\|f(t) - f_0(t)\| \leq |t^{-\alpha}| \left| \int_b^t v^{\alpha-1} \varphi(v) dv \right|$$

for any $t \in I$. Uniqueness is same as in case(a). This completes the proof of theorem.

Theorem 42 *Let X be a Complex Banach space and let $I = (a, b)$ be an open interval such that $0 < a < b \leq \infty$ or $-\infty < a < b < 0$. Assume that a function $\phi : I \to [0, \infty)$ is given and $h : I \to X$ is a continuous function. Furthermore, suppose a continuously differentiable function $f : I \to X$ satisfies*

$$\left\| ty'(t) + \alpha y(t) + h(t) \right\| \leq \phi(t), \qquad t \in I.$$

If both $t^{\alpha-1}\phi(t)$ and $t^{\alpha-1}h(t)$ are integrable on (a, c), for any c with $a < c \leq b$, then there exists a unique solution $f_0 : I \to X$ of the differential equation

$$ty'(t) + \alpha y(t) + h(t) = 0 \tag{5.8}$$

such that (5.3) holds for any $t \in I$, where α is a complex constant.

Proof 73 *The proof of the theorem follows from the proof of Theorem 41, if we set*

$$z(t) = \left(\frac{t}{a}\right)^\alpha f(t) + \int_a^t \frac{u^{\alpha-1}}{a^\alpha} h(u) du,$$

where the unique solution is given by

$$f_0(t) = \left(\frac{a}{t}\right)^\alpha - \frac{1}{t^\alpha} \int_a^t u^{\alpha-1} h(u) du.$$

The details are omitted.

5.2 Hyers-Ulam Stability of $t^2 y''(t) + \alpha t y'(t) + \beta y(t) = 0$

Let $I = (a, b)$ be an open interval with $0 < a < b \leq \infty$ or $-\infty < a < b < 0$. Assume that α, β are real constants satisfying either $\beta < 0$ or $\beta > 0$, $\alpha < 1$

and $(1-\alpha)^2 - 4\beta > 0$. Consider,

$$c = \frac{\alpha - 1 - \sqrt{(1-\alpha)^2 - 4\beta}}{2}, \qquad d = \frac{\alpha - 1 + \sqrt{(1-\alpha)^2 - 4\beta}}{2}.$$

Then the function

$$y(t) = \frac{c_1}{t^c} + \frac{c_2}{t^d}, \tag{5.9}$$

is the general solution of the Euler differential equation

$$t^2 y''(t) + \alpha t y'(t) + \beta y(t) = 0, \tag{5.10}$$

where c_1 and c_2 are arbitrary real numbers.

Theorem 43 *If a twice continuously differentiable function $f : I \to \mathbb{R}$ satisfies the differential inequality*

$$| t^2 f''(t) + \alpha t f'(t) + \beta f(t) | \leq \epsilon \tag{5.11}$$

for all $t \in I$ and for some $\epsilon > 0$, then there exists a solution $f_0 : I \to \mathbb{R}$ of the Euler equation (5.10) such that

$$|f(t) - f_0(t)| \leq \frac{\epsilon}{|\beta|} \left| \left(\frac{b}{t}\right)^c - 1 \right|$$

for any $t \in I$. In particular, if $I = (a, \infty)$ with $a > 0$, then

$$|f(t) - f_0(t)| \leq \frac{\epsilon}{|\beta|}$$

for all $t > a$.

Proof 74 *Define $g(t) = tf'(t) + cf(t)$ and $g'(t) = tf''(t) + f'(t) + cf'(t)$ for any $t \in I$, then*

$$\begin{aligned} |tg'(t) + dg(t)| &= |t\{tf''(t) + f'(t) + cf'(t)\} + d\{tf'(t) + cf(t)\}| \\ &= |t^2 f''(t) + tf'(t) + ctf'(t) + tdf'(t) + cdf(t)| \\ &= |t^2 f''(t) + t(1 + c + d)f'(t) + cdf(t)| \\ &= |t^2 f''(t) + \alpha t f'(t) + \beta f(t)| \\ &\leq \epsilon, \end{aligned} \tag{5.12}$$

for all $t \in I$, if $h(t) = -dg(t)$ and $h'(t) = -dg'(t)$, then inequality (5.12) becomes

$$\left| \frac{t}{d} h'(t) - h(t) \right| \leq \epsilon.$$

By using Theorem 2.3.3, there exists a real number c_0 such that

$$\left| h(t) - c_0 e^{d \int_a^t \frac{dT}{T}} \right| \leq \epsilon$$

implies that

$$\left|-dg(t) - c_0 e^{d[\ln T]_a^t}\right| \leq \epsilon,$$

that is,

$$\left|-d\{tf'(t) + cf(t)\} - c_0 e^{\ln\left(\frac{t}{a}\right)^d}\right| \leq \epsilon.$$

Hence,

$$\left|tf'(t) + cf(t) + c_0 \frac{a^d}{d} t^{-d}\right| \leq \frac{\epsilon}{|d|}$$

for every $t \in I$*, where* $c < 0$ *and* $c < d$*. Consider*

$$z(t) = \left(\frac{t}{a}\right)^c f(t) + \frac{c_0}{d(c-d)a^c}(t^{c-d} - a^{c-d})a^d$$

for all $t \in I$*. Proceeding as in Theorem 41,*

$$\|z(t) - z(s)\| \leq \left|\int_s^t \frac{v^{c-1}}{a^c} \frac{\epsilon}{d} dv\right|$$

for all $s, t \in I$*. By Theorem 41, there exists a solution* $f_0 : I \to \mathbb{R}$ *of the differential equation,*

$$ty'(t) + cy(t) + \frac{c_0 a^d}{d} t^{-d} = 0$$

such that

$$\|f(t) - f_0(t)\| \leq \frac{\epsilon}{|\beta|}\left|\left(\frac{b}{t}\right)^c - 1\right|$$

for any $t \in I$*. Indeed due to* (5.6)*, there exists a real number* c_3 *such that* f_0 *has the following form*

$$f_0(t) = \left(c_3 + \frac{c_0}{d(c-d)}\right)\frac{a^c}{t^c} - \frac{c_0}{d(c-d)}\frac{a^d}{t^d}.$$

Hence, $f_0(t)$ *is a solution of the Euler equation* (5.10)*.*

Example 10 *Consider the equation*

$$t^2 f''(t) - 6tf'(t) + 10f(t) = 0. \tag{5.13}$$

The function $f : I \to \mathbb{R}$ *satisfies the differential inequality*

$$|t^2 f''(t) - 6tf'(t) + 10f(t)| \leq \epsilon.$$

Assume that $f(t) = t^r$, $f'(t) = rt^{r-1}$ *and* $f''(t) = r(r-1)t^{r-2}$. *Then* (5.13) *becomes*

$$\begin{aligned} 0 &= t^2 r(r-1)t^{r-2} - t6rt^{r-1} + 10t^r \\ &= r(r-1)t^r - 6rt^r + 10t^r \\ &= (r^2 - r - 6r + 10)t^r \\ &= (r^2 - 7r + 10)t^2 \end{aligned}$$

implies that $r^2 - 7r + 10 = 0$, *that is,* $(r-2)(r-5) = 0$. *Let's define* $g(t) = tf'(t) - 2f(t)$ *and* $g'(t) = tf''(t) + f'(t) - 2f'(t)$, *then*

$$\begin{aligned} |tg'(t) - 5g(t)| &= |t\{tf''(t) + f'(t) - 2f'(t)\} - 5\{tf'(t) - 2f(t)\}| \\ &= |t^2 f''(t) + tf'(t) - 2tf'(t) - 5tf'(t) + 10f(t)| \\ &= |t^2 f''(t) + t(1-2-5)f'(t) + (-2. - 5)f(t)| \\ &= |t^2 - 6f'(t) + 10f(t)| \\ &\leq \epsilon. \end{aligned}$$

Using Theorem 41, there exists a solution $f_0 : I \to \mathbb{R}$ *of the differential equation* (5.13), *which is of the form*

$$\begin{aligned} f_0(t) &= \left(c_3 + \frac{c_0}{d(c-d)}\right)\frac{a^c}{t^c} - \frac{c_0}{d(c-d)}\frac{a^d}{t^d} \\ &= \left(c_3 - \frac{c_0}{35}\right)a^{-2}t^5 + \frac{c_0}{35}a^{-5}t^5. \end{aligned}$$

Hence $f_0(t)$ *is the solution of the Euler equation* (5.13).

5.3 Hyers-Ulam Stability of $t^3y''' + \alpha t^2y'' + \beta ty' + \gamma y = 0$

The objective of this section is to investigate the generalized Hyers-Ulam stability of the following Euler's differential equations of the form:

$$t^2y''(t) + \alpha ty'(t) + \beta y(t) + \gamma t^r x_0 = 0 \tag{5.14}$$

and

$$t^3y'''(t) + \alpha t^2y''(t) + \beta ty'(t) + \gamma y(t) = 0, \tag{5.15}$$

where α, β, γ and r are complex constants. Intuitively, we shall prove that if a twice continuously differentiable function $f : I \to X$ satisfies the differential inequality

$$\left\|t^2y''(t) + \alpha ty'(t) + \beta y(t) + \gamma t^r x_0\right\| \leq \phi(t) \tag{5.16}$$

for all $t \in I$, where x_0 is fixed element of the Complex Banach space X and $I = (a, b)$ with $0 < a < b \leq \infty$ or $-\infty < a < b < 0$, then there exists a twice continuously differentiable solution $f_0(t)$ of (5.14) such that

$$\|f(t) - f_0(t)\| \leq |t^l| \left| \int_t^b u^{-l-1} \psi(u) du \right| \tag{5.17}$$

for any $t \in I$, where

$$\psi(t) = |t^m| \left| \int_t^b u^{-m-1} \phi(u) du \right|.$$

We also apply this result to the investigation of the Hyers-Ulam stability of (5.15).

Let $I = (a, b)$ with $0 < a < b \leq \infty$ or $-\infty < a < b < 0$. For fixed $x_0 \neq 0$, the general solution of (5.14) in the class of real valued functions defined on I is given by

$$y(t) = \begin{cases} c_1 t^l + c_2 t^m - \frac{\gamma x_0 t^r}{g(r)}, & g(r) \neq 0, l \neq m, \\ c_1 t^l + c_2 t^m - \frac{\gamma x_0 t^r ln|t|}{g'(r)}, & g(r) = 0 \neq g'(r), l \neq m, \\ (c_1 + c_2 ln\,|t|)\, t^l - \frac{\gamma x_0 t^r}{g(r)}, & g(r) \neq 0, l = m, \\ (c_1 + c_2 ln\,|t|)\, t^l - \frac{\gamma x_0 t^r (ln|t|)^2}{2}, & g(r) = 0 = g'(r), l = m, \end{cases}$$

where $g(r) = r^2 + (\alpha - 1)r + \beta$, and c_1, c_2 are arbitrary constants.

Remark 6 *Indeed, $g'(r) = 2r + \alpha - 1$. If $g(r) = 0$, then either $r - l = 0$ or $r - m = 0$. Therefore, $r - l = 0$ implies that $2r + \alpha - 1 = 2r - l - m = r - m \neq 0$ and $r - m = 0$ implies that $2r + \alpha - 1 = r - l \neq 0$. Hence, the second solution could be any one of the following:*

$$c_1 t^l + c_2 t^m - \frac{\gamma x_0 t^r ln\,|t|}{r - m}; \quad r - l = 0, \quad r - m \neq 0, \; l \neq m,$$

$$c_1 t^l + c_2 t^m - \frac{\gamma x_0 t^r ln\,|t|}{r - l}; \quad r - l \neq 0, \quad r - m = 0, \; l \neq m.$$

Theorem 44 *Let X be a Complex Banach space and let $I = (a, b)$ be an open interval with either $0 < a < b \leq \infty$ or $-\infty < a < b < 0$. Assume that a function $\phi : I \to [0, \infty)$ is given, that α, β, γ, r are complex constants and that x_0 is a fixed element of X . Furthermore, suppose a twice continuously differentiable function $f : I \to X$ satisfies the differential inequality (5.16). If $t^{r-l-1}, t^{r-m-1}, t^{m-l-1}, t^{-m-1}\phi(t)$ and $t^{-l-1}\psi(t)$ are integrable on (a, c), for any c with $a < c \leq b$, then there exists a unique solution $f_0 : I \to X$ (which is twice continuously differentiable) of (5.14) such that (5.17) holds for any $t \in I$, where l and m are the roots of $g(r) = 0$.*

Proof 75 *We prove the theorem for five possible cases, viz.,*
(i) $(r-l)(r-m) \neq 0, \quad l \neq m;$
(ii) $r - l = 0 \neq r - m, \quad l \neq m;$
(iii) $r - l \neq 0 = r - m, \quad l \neq m;$
(iv) $r - l \neq 0, \qquad l = m;$
(v) $r - l = 0, \qquad l = m.$

Case *(i)* *Suppose that* X *is a Complex Banach space and* $f : I \to X$ *is a twice continuously differentiable function satisfying the differential inequality* (5.16). *Define* $h : I \to X$ *such that* $h(t) = tf'(t) - lf(t)$. *Then it is easy to verify that*

$$\begin{aligned}&\left\|t\ h'(t) - m\ h(t) + \gamma\ t^r x_0\right\| \\ &= \left\|t^2\ f''(t) + (1-l-m)t\ f'(t) + lm\ f(t) + \gamma\ t^r x_0\right\| \\ &= \left\|t^2\ f''(t) + \alpha t\ f'(t) + \beta\ f(t) + \gamma\ t^r x_0\right\| \leq\ \phi(t).\end{aligned}$$

Hence by Theorem 41, it follows that there exists a unique solution $h_0 : I \to X$ *of the differential equation* $ty' - my + \gamma\ t^r x_0 = 0$ *such that*

$$\|h(t) - h_0(t)\| \leq |t^m| \left|\int_t^b v^{-m-1}\phi(v)dv\right| \tag{5.18}$$

provided t^{r-m-1} *and* $t^{-m-1}\phi(t)$ *are integrable on* (a, c), *for any* c *with* $a < c \leq b$. *Indeed,*

$$h_0(t) = \left(\frac{a}{t}\right)^{-m} x - \frac{\gamma t^m}{r-m}\left(t^{r-m} - a^{r-m}\right) x_0,$$

where x *is a limit point in* X. *If we denote*

$$\psi(t) = \left|t^m \int_t^b v^{-m-1}\phi(v)dv\right|,$$

then (5.18) *becomes*

$$\left\|tf'(t) - lf(t) - h_0(t)\right\| \leq \psi(t). \tag{5.19}$$

Clearly, $\psi : I \to [0, \infty)$. *By Theorem 42, there exists a unique solution* $f_1 : I \to X$ *of the differential equation* $ty'(t) - ly(t) - h_0(t) = 0$ *such that*

$$\|f(t) - f_1(t)\| \leq |t^l| \left|\int_t^b u^{-l-1}\psi(u)du\right| \tag{5.20}$$

provided $t^{-l-1}\psi(t)$ *and* $t^{-l-1}h_0(t)$ *are integrable on* (a, c), *for any* c *with* $a <$

$c \leq b$. Clearly, $t^{-l-1}h_0(t)$ is integrable if and only if t^{m-l-1} and t^{r-l-1} are integrable. According to Theorem 42

$$f_1(t) = \left(\frac{t}{a}\right)^l x_1 + t^l \int_a^t u^{-l-1}h_0(u)du, \tag{5.21}$$

where x_1 is limit point in X. It is easy to verify that

$$\int_a^t u^{-l-1}h_0(u)du$$

$$= \left[\frac{x}{a^m} + \frac{\gamma x_0 a^{r-m}}{r-m}\right]\frac{t^{m-l}}{m-l} - \frac{x}{a^l(m-l)} - \frac{\gamma x_0 t^{r-l}}{(r-l)(r-m)} - \frac{\gamma x_0 a^{r-l}}{(r-l)(m-l)}.$$

Consequently,

$$\begin{aligned} f_1(t) = &\left[\frac{x}{a^l} - \frac{\gamma x_0 a^{r-l}}{(r-l)(m-l)} - \frac{x}{a^l(m-l)}\right]t^l \\ &+ \left[\frac{x}{a^m(m-l)} + \frac{\gamma x_0 a^{r-m}}{(r-m)(m-l)}\right]t^m - \frac{\gamma x_0 t^r}{g(r)} \end{aligned}$$

which is a solution of (5.14).

Case(ii) *Proceeding as in Case(i), we find a unique solution $f_2 : I \to X$ of the equation $ty'(t) - ly(t) - h_0(t) = 0$ such that*

$$\|f(t) - f_2(t)\| \leq |t^l| \left|\int_t^b u^{-l-1}\psi(u)du\right|$$

and

$$f_2(t) = \left(\frac{t}{a}\right)^l x_2 + t^l \int_a^t u^{-l-1}h_0(u)du,$$

where x_2 is a limit point in X. Clearly,

$$\begin{aligned} \int_a^t u^{-l-1}h_0(u)du &= \int_a^t \left[\frac{x}{a^m}u^{m-l-1} - \frac{\gamma x_0 u^{-1}}{r-m} + \frac{\gamma x_0 a^{r-m}}{r-m}u^{m-l-1}\right]du \\ &= \left[\frac{x}{a^m} + \frac{\gamma x_0 a^{r-m}}{r-m}\right]\frac{t^{m-l}}{m-l} \\ &+ \frac{\gamma x_0}{(r-m)(l-m)} - \frac{x}{a^l(m-l)} - \frac{\gamma x_0}{(r-m)} ln\left|\frac{t}{a}\right| \end{aligned}$$

implies that

$$\begin{aligned} f_2(t) = &\left[\frac{x_2}{a^l} + \frac{\gamma x_0}{(r-m)^2} + \frac{x}{a^l(r-m)}\right]t^l \\ &+ \left[\frac{x}{a^m(m-l)} - \frac{\gamma x_0 a^{r-m}}{(r-m)^2}\right]t^m - \frac{\gamma x_0 t^r}{(r-m)} ln\left|\frac{t}{a}\right| \end{aligned}$$

which is a solution of (5.14).

Case (iii) *Interchanging the role of l and m in Case (ii), we find the unique solution $f_3(t)$ as*

$$f_3(t) = \left[\frac{x_3}{a^l} - \frac{x}{a^l(m-l)} - \frac{\gamma x_0 a^{m-l}}{(m-l)^2}\right] t^l + \left[\frac{x}{a^m(m-l)} + \frac{\gamma x_0}{(m-l)^2}\right] t^m - \frac{\gamma x_0 t^r}{(r-l)} ln\left|\frac{t}{a}\right|,$$

where $x_3 \in X$.

Case (iv) *In this case, we proceed as in Case (i). Indeed, for $l = m$*

$$\int_a^t u^{-l-1} h_0(u) du = \left[\frac{x}{a^l} + \frac{\gamma x_0 a^{r-l}}{r-l}\right] ln\left|\frac{t}{a}\right| + \frac{\gamma x_0}{(r-l)^2}\left[a^{r-l} - t^{r-l}\right]$$

and the unique solution $f_4(t)$ is given by

$$f_4(t) = \left[\left\{\frac{x_4}{a^l} + \frac{\gamma x_0 a^{r-l}}{(r-l)^2}\right\} + \left\{\frac{x}{a^l} + \frac{\gamma x_0 a^{r-l}}{(r-l)}\right\} ln\left|\frac{t}{a}\right|\right] t^l - \frac{\gamma x_0 t^r}{(r-l)^2},$$

where $x_4 \in X$.

Case(v) *In this case,*

$$h_0(t) = \left(\frac{t}{a}\right)^l x - \gamma x_0 t^l ln\left|\frac{t}{a}\right|.$$

Using the same type of argument as in Case (i), we find the unique solution $f_5(t)$ as

$$\begin{aligned} f_5(t) &= \left(\frac{t}{a}\right)^l x_5 + t^l \int_a^t u^{-l-1} h_0(u) du \\ &= \left(\frac{t}{a}\right)^l x_5 + \frac{x}{a^l} t^l ln\left|\frac{t}{a}\right| - \frac{\gamma x_0}{2} t^l \left(ln\left|\frac{t}{a}\right|\right)^2 \\ &= \left[\frac{x_5}{a^l} + \frac{x}{a^l} ln\left|\frac{t}{a}\right|\right] t^l - \frac{\gamma x_0}{2} t^r \left(ln\left|\frac{t}{a}\right|\right)^2, \end{aligned}$$

where $x_5 \in X$. This completes the proof of the theorem.

Let $I = (a, b)$ with either $0 < a < b \leq \infty$ or $-\infty < a < b < 0$. We may note that $\alpha = 3-(l+m+n)$, $\beta = lm+mn+nl-l-m-n+1$ and $\gamma = -lmn$, when l, m and n are the characteristic roots of the associated characteristic equation of (5.15).

Theorem 45 *Let X be a Complex Banach space. Assume that $\theta : I \to [0, \infty)$ is given. Furthermore, assume that $t^{n-l-1}, t^{n-m-1}, t^{m-l-1}, t^{-n-1}\theta(t)$, $t^{-m-1}\lambda(t)$ and $t^{-l-1}\eta(t)$ are integrable on (a, c) with $a < c \leq b$, where*

$$\lambda(t) = \left|t^n \int_t^b u^{-n-1}\theta(u) du\right|, \quad \eta(t) = \left|t^m \int_t^b v^{-m-1}\lambda(v) dv\right|.$$

Suppose that $f \in C^3(I, X)$ *and satisfies*

$$\left\|t^3 f^{'''}(t) + \alpha t^2 f^{''}(t) + \beta t f^{'}(t) + \gamma f(t)\right\| \leq \theta(t), \tag{5.22}$$

for all $t \in I$. *Then there exists a unique solution* $f_0 \in C^3(I, X)$ *of* (5.15) *such that*

$$\|f(t) - f_0(t)\| \leq \left| t^l \int_t^b s^{-l-1} \eta(s) ds \right|.$$

Proof 76 *Let* $f : I \to X$ *be such that (3.1) hold, for* $t \in I$. *Define* $w : I \to X$ *such that*

$$w(t) = t^2 f^{''}(t) + (1 - l - m) t f^{'}(t) + lmf(t).$$

Indeed,

$$\left\|t w^{'}(t) - n w(t)\right\| \leq \theta(t), \quad t \in I.$$

Hence, using Theorem 42 it follows that there exists a unique $w_0 : I \to X$ *such that*

$$\|w(t) - w_0(t)\| \leq |t^n| \left| \int_t^b u^{-n-1} \theta(u) du \right| = \lambda(t), \tag{5.23}$$

where $w_0(t) = \left(\frac{t}{a}\right)^n x$ *and* x *is a limit point in* X. *Ultimately,* (5.23) *becomes*

$$\left\|t^2 f^{''}(t) + (1 - l - m) t f^{'}(t) + lmf(t) - a^{-n} t^n x\right\| \leq \lambda(t), \quad t \in I.$$

Now, we can apply Theorem 44 to the above differential inequality for five possible cases. Consequently,

$$\begin{aligned} f_1(t) &= \left[x_1 - \frac{x_2}{m - l} + \frac{x}{(n - l)(n - m)}\right] \left(\frac{t}{a}\right)^l \\ &+ \left[\frac{x_2}{m - l} + \frac{x}{(n - m)(l - m)}\right] \left(\frac{t}{a}\right)^m + \frac{x}{(n - l)(n - m)} \left(\frac{t}{a}\right)^n, \end{aligned}$$

$$\begin{aligned} f_2(t) = &\left[\frac{x_3}{a^l} + \frac{x_2}{a^l (n - m)} - \frac{x}{a^n (n - m)^2} + \frac{x}{a^n (n - m)} ln \left|\frac{t}{a}\right|\right] t^l \\ &+ \left[\frac{x_2}{a^m (m - l)} + \frac{x}{a^m (n - m)^2}\right] t^m, \end{aligned}$$

$$\begin{aligned} f_3(t) = &\left[\frac{x_4}{a^l} - \frac{x_2}{a^l (m - l)} + \frac{x}{a^l (m - l)^2}\right] t^l \\ &+ \left[\frac{x_2}{a^m (m - l)} - \frac{x}{a^m (m - l)^2} + \frac{x}{a^m (m - l)} ln \left|\frac{t}{a}\right|\right] t^m, \end{aligned}$$

$$f_4(t) = \left[\frac{x_4}{a^l} - \frac{x}{a^l(n-l)^2} + \left\{\frac{x_2}{a^l} - \frac{x}{a^l(n-l)}\right\} ln\left|\frac{t}{a}\right|\right] t^l + \frac{x}{(n-l)^2}\left(\frac{t}{a}\right)^n$$

and

$$f_5(t) = \left[\frac{x_5}{a^l} + \frac{x_2}{a^l} ln\left|\frac{t}{a}\right| + \frac{x}{2a^n}\left(ln\left|\frac{t}{a}\right|\right)^2\right] t^l$$

are the unique solutions with respect to the five possible cases respectively. In fact, they are the possible solutions of (5.15)*. Hence, the theorem is proved.*

Corollary 26 *Let* $X = \mathbb{R}$ *be a real Banach space. Assume that* t^{n-l-1}, t^{n-m-1} *and* t^{m-l-1} *are integrable on* $I = (a, b)$ *with* $0 < a < b \leq \infty$ *or* $-\infty < a < b < 0$, *where* l, m *and* n *are positive roots of the associated characteristic equation of* (5.15)*. Suppose that* $f \in C^3(I, X)$ *and satisfies*

$$\left\|t^3 f'''(t) + \alpha t^2 f''(t) + \beta t f'(t) + \gamma f(t)\right\| \leq \epsilon, \tag{5.24}$$

for all $t \in I$. *Then, there exists a unique solution* $f_0 \in C^3(I, X)$ *of* (5.15) *such that*

$$\|f(t) - f_0(t)\| \leq \frac{|\epsilon|}{lmn}.$$

Proof 77 *Let* $f : I \to X$ *be such that* (5.24) *hold, for all* $t \in I$. *Since all conditions of the Theorem 45 are satisfied, then there exists a unique solution* $f_0 : I \to X$ *of* (5.15) *such that*

$$\lambda(t) = \left|t^n \int_t^b u^{-n-1}\epsilon du\right| = |\epsilon|\,|t^n| \left|\left[\frac{u^{-n}}{-n}\right]_t^b\right| = \left|\frac{\epsilon}{n}\right|$$

and

$$\eta(t) = \left|t^m \int_t^b v^{-m-1}\lambda(v)dv\right| = \left|\frac{\epsilon}{mn}\right|\left|\left(\frac{t}{b}\right)^m - 1\right| = \left|\frac{\epsilon}{mn}\right|.$$

when $I = (-\infty, \infty)$. *Hence,*

$$\|f(t) - f_0(t)\| \leq \left|t^l \int_t^b s^{-l-1}\eta(s)ds\right| = \left|\frac{\epsilon}{lmn}\right|\left|\left(\frac{t}{b}\right)^l - 1\right|.$$

Consequently,

$$\|f(t) - f_0(t)\| \leq \frac{|\epsilon|}{lmn} \quad as\ b \to \infty,$$

where $f_0(t)$ *can be chosen from Theorem 45.*

Example 11 *Let $X = \mathbb{R}$ be a Banach space and $I = (a, \infty)$, $a > 0$. Consider the Euler's equation*

$$t^3 y^{'''}(t) - 6t^2 y^{''}(t) + 18t y^{'}(t) - 24y(t) = 0 \tag{5.25}$$

with $l = 4, m = 3$ and $n = 2$. Suppose that $f : I \to X$ satisfies the differential inequality

$$\left\| t^3 f^{'''}(t) - 6t^2 f^{''}(t) + 18t f^{'}(t) - 24f(t) \right\| \leq \epsilon$$

for any $t \in I$. By Corollary 26, there exists a unique solution $f_0 : I \to X$ of (5.25) such that

$$\|f(t) - f_0(t)\| \leq \frac{\epsilon}{24}$$

for any $t \in I$ and for unique $x, x_1, x_2 \in X$,

$$f_0(t) = \left\{\frac{x}{2} + x_1 + x_2\right\}\left(\frac{t}{a}\right)^4 + \{x - x_2\}\left(\frac{t}{a}\right)^3 + \left(\frac{x}{2}\right)\left(\frac{t}{a}\right)^2.$$

5.4 Hyers-Ulam Stability of $t^4 y^{(iv)} + \alpha t^3 y^{'''} + \beta t^2 y^{''} + \gamma t y^{'} + \delta y = 0$

The aim of this section is to investigate the generalized Hyers-Ulam stability of the following Euler's differential equations of the form:

$$t^3 y^{'''}(t) + \alpha t^2 y^{''}(t) + \beta t y^{'}(t) + \gamma y(t) + \delta t^r x_0 = 0 \tag{5.26}$$

and

$$t^4 y^{(iv)}(t) + \alpha t^3 y^{'''}(t) + \beta t^2 y^{''}(t) + \gamma t y(t) + \delta y(t) = 0, \tag{5.27}$$

where $\alpha, \beta, \gamma, \delta$ and r are complex constants with $x_0(\neq 0) \in X$. Here, we prove that if a function $f \in C^3(I, X)$ satisfies the differential inequality

$$\left\| t^3 y^{'''}(t) + \alpha t^2 y^{''}(t) + \beta t y^{'}(t) + \gamma y(t) + \delta t^r x_0 \right\| \leq \phi(t) \tag{5.28}$$

for all $t \in I$, where $x_0 \in X$ be a fixed element and $I = (a, b)$ with $0 < a < b \leq \infty$ or $-\infty < a < b < 0$, then there exists a unique solution $f_0 \in C^3(I, X)$ of (5.28) such that

$$\|f(t) - f_0(t)\| \leq |t^n| \left| \int_t^b u^{-n-1} \theta(u) du \right| \tag{5.29}$$

for all $t \in I$, where

$$\theta(t) = |t^l| \left| \int_t^b u^{-l-1} \psi(u) du \right|, \quad \psi(t) = |t^m| \left| \int_t^b v^{-m-1} \phi(v) dv \right| \tag{5.30}$$

and l, m, n are the characteristic roots of (5.26) such that $\alpha = 3 - (l+m+n)$, $\beta = lm + mn + nl - l - m - n + 1$ and $\gamma = -(lmn)$. Also, we apply this result to investigate the Hyers-Ulam stability of (5.27). For this, we let $I = (a, b)$ either $0 < a < b \leq \infty$ or $-\infty < a < b < 0$.

Let l, m and n be the characteristic roots of (5.26) such that $\alpha = 3 - (l + m + n)$, $\beta = lm + mn + nl - l - m - n + 1$ and $\gamma = -(lmn)$ with $g(r) = r^3 + (\alpha - 3)r^2 + (\beta - \alpha + 2)r + \gamma$. For fixed $x_0 \neq 0$, the possible solutions of (5.26) in the class of real valued functions defined on I are given by

(I) when $g(r) \neq 0$,

$$y(t) = \begin{cases} c_1 t^l + c_2 t^m + c_3 t^n - \frac{\delta x_0 t^r}{g(r)}, & l \neq m \neq n, \\ c_1 t^l + (c_2 + c_3 ln\,|t|) t^m - \frac{\delta x_0 t^r}{g(r)}, & l \neq m = n, \\ (c_1 + c_2 ln\,|t|)\, t^l + c_3 t^n - \frac{\delta x_0 t^r}{g(r)}, & l = m \neq n, \\ (c_1 + c_3 ln\,|t|)\, t^n + c_2 t^m - \frac{\delta x_0 t^r}{g(r)}, & l = n \neq m, \\ \left\{ c_1 + c_2 ln\,|t| + c_3 (ln\,|t|)^2 \right\} t^l - \frac{\delta x_0 t^r}{g(r)}, & l = m = n; \end{cases}$$

(II) when $g(r) = 0 \neq g'(r)$,

$$y(t) = \begin{cases} c_1 t^l + c_2 t^m + c_3 t^n - \frac{\delta x_0 t^r ln|t|}{g'(r)}, & l \neq m \neq n, \\ c_1 t^l + (c_2 + c_3 ln\,|t|) t^m - \frac{\delta x_0 t^r ln|t|}{g'(r)}, & r - l = 0, l \neq m = n, \\ (c_1 + c_2 ln\,|t|)\, t^l + c_3 t^n - \frac{\delta x_0 t^r ln|t|}{g'(r)}, & r - n = 0, l = m \neq n, \\ (c_1 + c_3 ln\,|t|)\, t^n + c_2 t^m - \frac{\delta x_0 t^r ln|t|}{g'(r)}, & r - m = 0, l = n \neq m. \end{cases}$$

Remark 7 *Indeed,* $g'(r) = 3r^2 + 2(\alpha - 3)r + (\beta - \alpha + 2)$. *If* $g(r) = 0$, *then either* $r - l = 0$ *or* $r - m = 0$ *or* $r - n = 0$. *Therefore,* $r - l = 0$ *and* $g'(r) \neq 0$ *implies that* $3r^2 + 2(\alpha - 3)r + (\beta - \alpha + 2) = (r - m)(r - n) \neq 0$. *So* $r - m \neq 0$ *and* $r - n \neq 0$. *Similarly, when* $r - m = 0$ *and* $g'(r) \neq 0$ *implies* $r - l \neq 0$ *and*

$r-n \neq 0$ and when $r-n=0$ and $g'(r) \neq 0$ implies $r-l =\neq 0$ and $r-m \neq 0$. Hence, the first solution for $l \neq m \neq n$ could be any one of the following:

$$c_1 t^l + c_2 t^m + c_3 t^n - \frac{\delta x_0 t^r ln\,|t|}{(r-m)(r-n)}; \qquad r-l=0,\ r-m \neq 0,\ r-n \neq 0,$$

$$c_1 t^l + c_2 t^m + c_3 t^n - \frac{\delta x_0 t^r ln\,|t|}{(r-l)(r-m)}; \qquad r-n=0,\ r-l \neq 0,\ r-m \neq 0,$$

$$c_1 t^l + c_2 t^m + c_3 t^n - \frac{\delta x_0 t^r ln\,|t|}{(r-l)(r-n)}; \qquad r-m=0,\ r-l \neq 0,\ r-n \neq 0.$$

(III) When $g(r) = 0 = g'(r)$, but $g''(r) \neq 0$

$$y(t) = \begin{cases} (c_1 + c_2 ln\,|t|)\, t^l + c_3 t^n - \frac{\delta x_0 t^r (ln|t|)^2}{g''(r)}, & l=m=r, \quad r-n \neq 0, \\ c_1 t^l + (c_2 + c_3 ln\,|t|) t^m - \frac{\delta x_0 t^r (ln|t|)^2}{g''(r)}, & m=n=r, \quad r-l \neq 0, \\ (c_1 + c_3 ln\,|t|)\, t^n + c_2 t^m - \frac{\delta x_0 t^r (ln|t|)^2}{g''(r)}, & l=n=r, \quad r-m \neq 0. \end{cases}$$

(IV) When $g(r) = 0 = g'(r) = g''(r)$,

$$y(t) = \left\{c_1 + c_2 ln\,|t| + c_3\,(ln\,|t|)^2\right\} t^l - \frac{\delta x_0 t^r\,(ln\,|t|)^3}{g'''(r)},$$

where c_1, c_2 and c_3 are arbitrary constants.

Theorem 46 *Let X be a Complex Banach space. Assume that a function $\phi : I \to [0,\infty)$ is given, that $\alpha, \beta, \gamma, \delta, r$ are complex constants and that x_0 is a fixed element of X. Furthermore, suppose a thrice continuously differentiable function $f : I \to X$ satisfies the differential inequality (5.28). If $t^{r-l-1}, t^{m-l-1}, t^{l-n-1}, t^{m-n-1}, t^{r-n-1}, t^{m-l-1} ln\left|\frac{t}{a}\right|, t^{l-n-1} ln\left|\frac{t}{a}\right|, t^{r-n-1}$ $ln\left|\frac{t}{a}\right|, t^{r-n-1}\left(ln\left|\frac{t}{a}\right|\right)^2, t^{-m-1}\phi(t), t^{-l-1}\psi(t)$ and $t^{-n-1}\theta(t)$ are integrable on (a,c), for any c with $a < c \leq b$, then there exists a unique solution $f_0 \in C^3(I,X)$ of (5.26) such that (5.29) holds for any $t \in I$, where l, m and n are the roots of $g(r)=0$.*

Proof 78 *To prove the theorem it is sufficient to consider the following cases:*
(i) $(r-l)(r-m)(r-n) \neq 0,\ \ l \neq m \neq n;$
(ii) $(r-l)(r-m)(r-n) \neq 0,\ \ l \neq m = n;$
(iii) $(r-l)(r-m)(r-n) \neq 0,\ \ l = m \neq n;$
(iv) $(r-l)(r-m)(r-n) \neq 0,\ \ l = n \neq m;$
(v) $(r-l)(r-m)(r-n) \neq 0,\ \ l = m = n;$

(*vi*) $(r-l)=0,(r-m)\neq 0\neq(r-n),\quad l\neq m\neq n;$
(*vii*) $(r-n)=0,(r-m)\neq 0\neq(r-l),\quad l\neq m\neq n;$
(*viii*) $(r-m)=0,(r-l)\neq 0\neq(r-n),\quad l\neq m\neq n;$
(*ix*) $(r-l)=0,(r-m)\neq 0\neq(r-n),\quad l\neq m=n;$
(*x*) $(r-n)=0,(r-m)\neq 0\neq(r-l),\quad l=m\neq n;$
(*xi*) $(r-m)=0,(r-l)\neq 0\neq(r-n),\quad l=n\neq m;$
(*xii*) $(r-l)=0=(r-m),(r-n)\neq 0,\quad l=m\neq n;$
(*xiii*) $(r-m)=0=(r-n),(r-l)\neq 0,\quad l\neq m=n;$
(*xiv*) $(r-l)=0=(r-n),(r-m)\neq 0,\quad l=n\neq m;$
(*xv*) $(r-l)=(r-m)=(r-n)=0,\quad l=m=n.$

Case-(i) Suppose that X is a Complex Banach space and a thrice continuously differentiable function $f: I\to X$ is satisfying the differential inequality (5.26). *Let l, m and n be the roots of $g(u)=u^3+(\alpha-3)u^2+(\beta-\alpha+2)u+\gamma=0$. Define $h: I\to X$ such that $h(t)=tf'(t)-nf(t)$. Then we have*

$$\begin{aligned}&\left\|t^2\, h''(t)+(1-l-m)t\, h'(t)+lm\, h(t)+\delta\, t^r x_0\right\|\\&=\left\|t^3\, f'''(t)+\alpha\, t^2\, f''(t)+\beta\, t\, f'(t)+\gamma\, f(t)+\delta\, t^r x_0\right\|\\&\leq\ \phi(t).\end{aligned}$$

Hence by Theorem 44 and then using (5.30), *it follows that there exists a unique solution $h_0: I\to X$ of the differential equation*

$$t^2y''(t)+(1-l-m)ty'(t)+lmy(t)+\delta t^r x_0=0$$

such that

$$\|h(t)-h_0(t)\|\leq\left|t^l\right|\left|\int_t^b u^{-l-1}\psi(u)du\right|=\theta(t),\tag{5.31}$$

where $h_0(t)=k_1t^l+k_2t^m+k_3t^r$ with

$$\begin{aligned}k_1&=\left\{\frac{x_1}{a^l}-\frac{\delta x_0a^{r-l}}{(r-l)(m-l)}-\frac{x}{a^l(m-l)}\right\},\\k_2&=\left\{\frac{x}{a^m(m-l)}-\frac{\delta x_0a^{r-m}}{(r-m)(l-m)}\right\},\\k_3&=-\frac{\delta x_0}{(r-l)(r-m)}.\end{aligned}$$

Consequently, (5.31) *becomes*

$$\left\|tf'(t)-nf(t)-h_0(t)\right\|\leq\theta(t).\tag{5.32}$$

Clearly $\theta: I\to[0,\infty)$. By Theorem 42 there exists a unique solution $f_0: I\to$

X of the differential equation $ty'(t) - ny(t) - h_0(t) = 0$ such that

$$\|f(t) - f_0(t)\| \leq |t^n| \left| \int_t^b v^{-n-1}\theta(v)dv \right|, \qquad t \in I \tag{5.33}$$

provided $t^{-n-1}\theta(t)$ and $t^{-n-1}h_0(t)$ are integrable on (a,c), for any c with $a < c \leq b$. As $t^{l-n-1}, t^{m-n-1}, t^{r-n-1}, t^{l-n-1}ln\left|\frac{t}{a}\right|, t^{r-n-1}ln\left|\frac{t}{a}\right|$ and $t^{l-n-1}\left(ln\left|\frac{t}{a}\right|\right)^2$ are integrable, then so also $t^{-n-1}h_0(t)$. According to Theorem 42, $f_0(t)$ is given by

$$f_0(t) = \left(\frac{t}{a}\right)^n \bar{x} + t^n \int_a^t v^{-n-1}h_0(v)dv, \tag{5.34}$$

where $\bar{x}$ is a limit point in X. It is easy to verify that

$$\int_a^t v^{-n-1}h_0(v)dv = \frac{k_1}{l-n}\left(t^{l-n} - a^{l-n}\right) + \frac{k_2}{m-n}\left(t^{m-n} - a^{m-n}\right) + \frac{k_3}{r-n}\left(t^{r-n} - a^{r-n}\right).$$

As a result,

$$f_0(t) = \frac{k_1}{l-n}t^l + \frac{k_2}{m-n}t^m + \left\{\frac{\bar{x}}{a^n} - \frac{k_1 a^{l-n}}{l-n} - \frac{k_2 a^{m-n}}{m-n} - \frac{k_3 a^{r-n}}{r-n}\right\} t^n - \frac{\delta x_0 t^r}{g(r)}.$$

Case-(ii) Proceeding as in Case (i), we can obtain

$$\int_a^t v^{-n-1}h_0(v)dv = \frac{k_1}{l-n}\left(t^{l-n} - a^{l-n}\right) + k_2\ ln\left|\frac{t}{a}\right| + \frac{k_3}{r-n}\left(t^{r-n} - a^{r-n}\right).$$

Consequently,

$$f_0(t) = \frac{k_1}{l-m}t^l + \left\{\frac{\bar{x}}{a^m} - \frac{k_1 a^{l-m}}{l-m} - \frac{k_3 a^{r-m}}{r-m} + k_2\ ln\left|\frac{t}{a}\right|\right\} t^m - \frac{\delta x_0 t^r}{(r-l)(r-m)^2},$$

where k_1, k_2 and k_3 are same as in Case (i).
Case-(iii) Proceeding as before with

$$h_0(t) = \{k_4 + k_5 ln\left|\frac{t}{a}\right|\}t^l + k_6 t^r$$

and

$$k_4 = \left\{\frac{x_2}{a^l} + \frac{\delta x_0 a^{r-l}}{(r-l)^2}\right\}, \quad k_5 = \left\{\frac{x}{a^l} + \frac{\delta x_0 a^{r-l}}{r-l}\right\}, \quad k_6 = -\frac{\delta x_0}{(r-l)^2},$$

it is easy to verify that

$$\int_a^t v^{-n-1}h_0(v)dv = \frac{k_4}{l-n}\left(t^{l-n} - a^{l-n}\right)$$
$$+\frac{k_5}{l-n}\left[\left(ln\left|\frac{t}{a}\right| - \frac{1}{l-n}\right)t^{l-n} + \frac{a^{l-n}}{(l-n)^2}\right] + \frac{k_6}{r-n}\left(t^{r-n} - a^{r-n}\right).$$

Hence, from (5.34)

$$f_0(t) = \left[\frac{k_4}{l-n} - \frac{k_5}{(l-n)^2} + \frac{k_{11}}{l-n}ln\left|\frac{t}{a}\right|\right]t^l$$
$$+\left[\frac{\bar{x}}{a^n} - \frac{k_4a^{l-n}}{l-n} + \frac{k_5a^{l-n}}{(l-n)^2} - \frac{k_6a^{r-n}}{r-n}\right]t^n - \frac{\delta x_0 t^r}{(r-n)(r-l)^2}.$$

Case-(iv) We proceed as in Case (i) and it is easy to see that

$$\int_a^t v^{-n-1}h_0(v)dv = k_1 ln\left|\frac{t}{a}\right| + \frac{k_2}{m-n}\left(t^{m-n} - a^{m-n}\right) + \frac{k_3}{r-n}\left(t^{r-n} - a^{r-n}\right).$$

So from (5.34), *it happens that*

$$f_0(t) = \frac{k_2}{m-n}t^m + \left\{\frac{\bar{x}}{a^n} - \frac{k_2a^{m-n}}{m-n} - \frac{k_3a^{r-n}}{r-n} + k_1 ln\left|\frac{t}{a}\right|\right\}t^n - \frac{\delta x_0 t^r}{(r-m)(r-n)^2}.$$

Case-(v) In this case we have that

$$h_0(t) = \left\{k_4 + k_5 ln\left|\frac{t}{a}\right|\right\}t^l + k_6t^r,$$

where k_4, k_5 *and* k_6 *are same as in Case (iii). Therefore,*

$$\int_a^t v^{-n-1}h_0(v)dv = k_4\ ln\left|\frac{t}{a}\right| + \frac{k_5}{2}\left(ln\left|\frac{t}{a}\right|\right)^2 + \frac{k_6}{r-n}\left(t^{r-n} - a^{r-n}\right)$$

implies from (5.34) *that*

$$f_0(t) = \left[\frac{\bar{x}}{a^n} - \frac{k_6a^{r-l}}{r-l} + k_4 ln\left|\frac{t}{a}\right| + \frac{k_5}{2}\left(ln\left|\frac{t}{a}\right|\right)^2\right]t^l - \frac{\delta x_0 t^r}{(r-l)^3}.$$

Case-(vi) In this case, we notice that

$$h_0(t) = k_7t^l + k_8t^m + k_9t^r ln\left|\frac{t}{a}\right|,$$

where

$$k_7 = \left\{\frac{x_3}{a^l} + \frac{\delta x_0}{(r-m)^2} + \frac{x}{a^l(r-m)}\right\},$$
$$k_8 = \left\{\frac{x}{a^m(m-l)} - \frac{\delta x_0 a^{r-m}}{(r-m)^2}\right\},\quad k_9 = -\frac{\delta x_0}{(r-m)},$$

and

$$\int_a^t v^{-n-1}h_0(v)dv = \frac{k_7}{l-n}\left(t^{l-n}-a^{l-n}\right)+\frac{k_8}{m-n}\left(t^{m-n}-a^{m-n}\right)$$
$$+k_9\left[\frac{t^r}{r-n}ln\left|\frac{t}{a}\right|-\frac{t^{r-n}}{(r-n)^2}-\frac{a^{r-n}}{(r-n)^2}\right].$$

Hence from (5.34), *it follows that*

$$f_0(t)=\left[\frac{k_7}{l-n}-\frac{k_9}{(l-n)^2}\right]t^l+\frac{k_8}{m-n}t^m$$
$$+\left[\frac{\bar{x}}{a^n}-\frac{k_7\ a^{l-n}}{l-n}-\frac{k_8\ a^{m-n}}{m-n}+\frac{k_9\ a^{l-n}}{(l-n)^2}\right]t^n-\frac{\delta x_0 t^r}{(r-m)(r-n)}\ ln\left|\frac{t}{a}\right|.$$

Case-(vii) Here, we have

$$h_0(t)=k_1t^l+k_2t^m+k_3t^r,$$

and

$$\int_a^t v^{-n-1}h_0(v)dv=\frac{k_1}{l-n}\left(t^{l-n}-a^{l-n}\right)+\frac{k_2}{m-n}\left(t^{m-n}-a^{m-n}\right)+k_3ln\left|\frac{t}{a}\right|,$$

where k_1, k_2 *and* k_3 *are same as in Case-*(i). *Applying this in* (5.34), *we obtain*

$$f_0(t)=\frac{k_1}{l-n}t^l+\frac{k_2}{m-n}t^m+\left[\frac{\bar{x}}{a^n}-\frac{k_1a^{l-n}}{l-n}-\frac{k_2a^{m-n}}{m-n}\right]t^n-\frac{\delta x_0 t^r}{(r-l)(r-m)}\ ln\left|\frac{t}{a}\right|.$$

Case-(viii) For this case, $h_0(t)$ *becomes*

$$h_0(t)=k_{10}t^l+k_{11}t^m+k_{12}t^r ln\left|\frac{t}{a}\right|,$$

where

$$k_{10}=\left\{\frac{x_4}{a^l}-\frac{\delta x_0 a^{m-l}}{(m-l)^2}-\frac{x}{a^l(m-l)}\right\},$$
$$k_{11}=\left\{\frac{x}{a^m(m-l)}+\frac{\delta x_0}{(m-l)^2}\right\},k_{12}=-\frac{\delta x_0}{(r-l)}.$$

Using

$$\int_a^t v^{-n-1}h_0(v)dv=\frac{k_{10}}{l-n}\left(t^{l-n}-a^{l-n}\right)+\frac{k_{11}}{m-n}\left(t^{m-n}-a^{m-n}\right)$$
$$+\frac{k_{12}}{r-n}\left[t^{r-n}ln\left|\frac{t}{a}\right|-\frac{t^{r-n}}{r-n}+\frac{a^{r-n}}{(r-n)^2}\right]$$

in (5.34), *we find*

$$f_0(t) = \left(\frac{k_{10}}{l-n}\right)t^l + \left\{\frac{k_{11}}{m-n} + \frac{\delta x_0}{(m-l)(m-n)^2}\right\}t^m$$
$$+ \left[\frac{\bar{x}}{a^n} - \frac{k_{10}\ a^{l-n}}{l-n} - \frac{k_{11}\ a^{m-n}}{m-n} - \frac{\delta x_0\ a^{m-n}}{(m-l)(m-n)^2}\right]t^n - \frac{\delta x_0 t^r}{(r-l)(r-n)}\ ln\left|\frac{t}{a}\right|.$$

Case-(ix) Here $h_0(t) = k_7 t^l + k_8 t^m + k_9 t^r ln\left|\frac{t}{a}\right|$, *where* k_7, k_8 *and* k_9 *are same as in Case-(vi). Using the fact*

$$\int_a^t v^{-n-1}h_0(v)dv$$
$$= \frac{k_7}{l-n}\left(t^{l-n} - a^{l-n}\right) + k_8\ ln\left|\frac{t}{a}\right| + \frac{k_9}{r-n}\left[t^{r-n}ln\left|\frac{t}{a}\right| - \frac{t^{r-n}}{r-n} + \frac{a^{r-n}}{r-n}\right]$$

in (5.34), *it follows that*

$$f_0(t) = \left(\frac{k_7}{l-m} - \frac{k_9}{(l-m)^2}\right)t^l$$
$$+ \left[\frac{\bar{x}}{a^m} - \frac{k_7\ a^{l-m}}{l-m} + \frac{k_9\ a^{l-m}}{(l-m)^2} + k_8 ln\left|\frac{t}{a}\right|\right]t^m - \frac{\delta x_0 t^r}{(r-m)^2}\ ln\left|\frac{t}{a}\right|.$$

Case-(x) In this case, $h_0(t) = \left\{k_4 + k_5 ln\left|\frac{t}{a}\right|\right\}t^l + k_6 t^r$, *where* k_4, k_5 *and* k_6 *are same as in Case-(iii). Now, it is easy to verify that*

$$\int_a^t v^{-n-1}h_0(v)dv = \frac{k_4}{l-n}\left(t^{l-n} - a^{l-n}\right)$$
$$+ \frac{k_5}{l-n}\left[t^{l-n}\ ln\left|\frac{t}{a}\right| - \frac{t^{l-n}}{l-n} + \frac{a^{l-n}}{l-n}\right] + k_6\ ln\left|\frac{t}{a}\right|.$$

Hence, from (5.34) *we obtain*

$$f_0(t) = \left[\left\{\frac{k_4}{l-n} - \frac{k_5}{(l-n)^2}\right\} + \frac{k_5}{l-n}\ ln\left|\frac{t}{a}\right|\right]t^l$$
$$+ \left[\frac{\bar{x}}{a^n} - \frac{k_4\ a^{l-n}}{l-n} + \frac{k_5\ a^{l-n}}{(l-n)^2}\right]t^n - \frac{\delta x_0 t^r}{(r-l)^2}\ ln\left|\frac{t}{a}\right|.$$

Case-(xi) In this case, $h_0(t) = k_{10}t^l + k_{11}t^m + k_{12}t^r ln\left|\frac{t}{a}\right|$, *where* k_{10}, k_{11} *and* k_{12} *are same as in Case-(viii). Clearly,*

$$\int_a^t v^{-n-1}h_0(v)dv = k_{10}\ ln\left|\frac{t}{a}\right| + \frac{k_{11}}{m-n}\left(t^{m-n} - a^{m-n}\right)$$
$$+ \frac{k_{12}}{r-n}\left[t^{r-n}\ ln\left|\frac{t}{a}\right| - \frac{t^{r-n}}{r-n} + \frac{a^{r-n}}{r-n}\right]$$

and (5.34) reduces to

$$f_0(t) = \left[\frac{\bar{x}}{a^l} - \frac{k_{11}\ a^{m-l}}{m-l} + \frac{k_{12}\ a^{m-l}}{(m-l)^2} + k_{10}\ ln\left|\frac{t}{a}\right|\right] t^l + \left[\frac{k_{11}}{m-l} - \frac{k_{12}}{(m-l)^2}\right] t^m - \frac{\delta x_0 t^r}{(r-n)^2}\ ln\left|\frac{t}{a}\right|.$$

Case-(xii) In this case we obtain

$$h_0(t) = \left\{\frac{x_5}{a^l} + \frac{x}{a^l}\ ln\left|\frac{t}{a}\right|\right\} t^l - \frac{\delta x_0 t^r}{2}\left(ln\left|\frac{t}{a}\right|\right)^2$$

and thus

$$\int_a^t v^{-n-1} h_0(v) dv$$
$$= \frac{x_5}{a^l(l-n)}\left[t^{l-n} - a^{l-n}\right] + \frac{x}{a^l(l-n)}\left[t^{l-n}\ ln\left|\frac{t}{a}\right| - \frac{t^{l-n}}{l-n} + \frac{a^{l-n}}{l-n}\right]$$
$$- \frac{\delta x_0}{2(r-n)}\left[t^{r-n}\left(ln\left|\frac{t}{a}\right|\right)^2 - \frac{2t^{r-n}}{r-n} ln\left|\frac{t}{a}\right| + \frac{2t^{r-n}}{(r-n)^2} - \frac{2a^{r-n}}{(r-n)^2}\right]$$

which on applying in (5.34), we get

$$f_0(t)$$
$$= \left[\frac{x_5}{a^l(l-n)} - \frac{x}{a^l(l-n)^2} - \frac{\delta x_0}{(l-n)^3} + \left\{\frac{x}{a^l(l-n)} + \frac{\delta x_0}{(l-n)^2}\right\} ln\left|\frac{t}{a}\right|\right] t^l$$
$$+ \left[\frac{\bar{x}}{a^n} - \frac{x_5}{a^n(l-n)} + \frac{x}{a^n(l-n)^2} + \frac{\delta x_0 a^{l-n}}{(l-n)^3}\right] t^n - \frac{\delta x_0 t^r}{2(r-n)}\left(ln\left|\frac{t}{a}\right|\right)^2.$$

Case-(xiii) Here $h_0(t) = k_{10}t^l + k_{11}t^m + k_{12}t^r ln\left|\frac{t}{a}\right|$, *where* k_{10}, k_{11} *and* k_{12} *are same as in Case-(viii) and*

$$\int_a^t v^{-n-1} h_0(v) dv = \frac{k_{10}}{l-n}\left(t^{l-n} - a^{l-n}\right) + k_{11}\ ln\left|\frac{t}{a}\right| + \frac{k_{12}}{2}\left(ln\left|\frac{t}{a}\right|\right)^2.$$

Applying this in (5.34), it follows that

$$f_0(t) = \frac{k_{10}}{l-n} t^l + \left\{\frac{\bar{x}}{a^n} - \frac{k_{10}\ a^{l-n}}{l-n} + k_{11}\ ln\left|\frac{t}{a}\right|\right\} t^n - \frac{\delta x_0 t^r}{2(r-l)}\left(ln\left|\frac{t}{a}\right|\right)^2.$$

Case-(xiv) In this case,

$$h_0(t) = k_7 t^l + k_8 t^m + k_9 t^r ln\left|\frac{t}{a}\right|,$$

where k_7, k_8 *and* k_9 *are same as in Case-(vi) and*

$$\int_a^t v^{-n-1}h_0(v)dv = k_7\ ln\left|\frac{t}{a}\right| + \frac{k_8}{m-n}\left(t^{m-n} - a^{m-n}\right) + \frac{k_9}{2}\left(ln\left|\frac{t}{a}\right|\right)^2.$$

Ultimately, (5.34) *becomes*

$$f_0(t) = \left(\frac{k_8}{m-n}\right)t^m + \left\{\frac{\bar{x}}{a^n} - \frac{k_8\ a^{m-n}}{m-n} + k_7\ ln\left|\frac{t}{a}\right|\right\}t^n - \frac{\delta x_0 t^r}{2(r-m)}\left(ln\left|\frac{t}{a}\right|\right)^2.$$

Case-(xv) For this case,

$$h_0(t) = \left\{\frac{x_5}{a^l} + \frac{x}{a^l}\ ln\left|\frac{t}{a}\right|\right\}t^l - \frac{\delta x_0 t^r}{2}\left(ln\left|\frac{t}{a}\right|\right)^2$$

and hence

$$\int_a^t v^{-n-1}h_0(v)dv = \frac{x_5}{a^l}\ ln\left|\frac{t}{a}\right| + \frac{x}{2a^l}\left(ln\left|\frac{t}{a}\right|\right)^2 - \frac{\delta x_0}{6}\left(ln\left|\frac{t}{a}\right|\right)^3.$$

Using this in (5.34), $f_0(t)$ *can be obtained as*

$$f_0(t) = \left[\frac{\bar{x}}{a^l} + \frac{x_5}{a^l}\ ln\left|\frac{t}{a}\right| + \frac{x}{2a^l}\left(ln\left|\frac{t}{a}\right|\right)^2\right]t^l - \frac{\delta x_0 t^r}{6}\left(ln\left|\frac{t}{a}\right|\right)^3.$$

Here $\bar{x}, x, x_1, x_2, x_3, x_4$ *and* x_5 *are all limit points in* X*. This completes the proof of the theorem.*

Assume that l, m, n and p are the characteristic roots of (5.27) such that

$$\begin{aligned}
&\alpha = \{6 - (l+m+n+p)\},\\
&\beta = \{7 - 3(l+m+n+p) + (lm+mn+nl+np+lp+mp)\},\\
&\gamma = 1 - (l+m+n+p) + (lm+mn+nl+mp+lp+np)\\
&- (lmn+mnp+lnp+lmp),\\
&\delta = lmnp,\ \alpha_1 = \{3-(l+m+n)\},\ \beta_1 = \{lm+mn+nl-l-m-n+1\}\\
&\gamma_1 = -(lmn).
\end{aligned}$$

Theorem 47 *Let* X *be a complex Banach space. Assume that a function* $\eta : I \to [0,\infty)$ *is given. Furthermore, assume that* $t^{p-l-1}, t^{m-l-1}, t^{l-n-1}, t^{m-n-1}, t^{p-n-1}, t^{m-l-1}ln\left|\frac{t}{a}\right|, t^{l-n-1}ln\left|\frac{t}{a}\right|, t^{p-n-1}ln\left|\frac{t}{a}\right|, t^{p-n-1}\left(ln\left|\frac{t}{a}\right|\right)^2, t^{-p-1}\eta(t)$, $t^{-m-1}\phi(t)$, $t^{-l-1}\psi(t)$ *and* $t^{-n-1}\theta(t)$ *are integrable over the interval* (a,c) *with* $a < c \leq b$*, where*

$$\phi(t) = |t^p|\left|\int_t^b v^{-p-1}\eta(v)dv\right|, \qquad \psi(t) = |t^m|\left|\int_t^b v^{-m-1}\phi(v)dv\right|,$$

$$\theta(t) = \left|t^l\right|\left|\int_t^b u^{-l-1}\psi(u)du\right|.$$

Suppose that $f \in C^4(I, X)$ satisfies the differential inequality

$$\left\| t^4 y^{(iv)}(t) + \alpha t^3 y'''(t) + \beta t^2 y''(t) + \gamma t y(t) + \delta y(t) \right\| \leq \eta(t) \tag{5.35}$$

for all $t \in I$. Then there exists a unique solution $f_0 \in C^4(I, X)$ of (5.27) such that

$$\|f(t) - f_0(t)\| \leq \left| t^n \int_t^b v^{-n-1} \theta(v) dv \right|.$$

Proof 79 *Let X be a complex Banach space and $f : I \to X$ such that (5.35) hold for $t \in I$. Define $s : I \to X$ such that*

$$s(t) = t^3 f'''(t) + \alpha_1 t^2 f''(t) + \beta_1 t f'(t) + \gamma_1 f(t). \tag{5.36}$$

Indeed,

$$\begin{aligned} \|ts'(t) - ps(t)\| &= \left\| t^4 f^{(iv)}(t) + \alpha t^3 f'''(t) + \beta t^2 f''(t) + \gamma t f'(t) + \delta f(t) \right\| \\ &\leq \eta(t). \end{aligned}$$

From Theorem 42, it follows that there exists a unique solution $s_0 : I \to X$ of the differential equation $ts'(t) - ps(t) = 0$ such that

$$\|s(t) - s_0(t)\| \leq |t^p| \left| \int_t^b u^{-p-1} \eta(u) du \right|,$$

where $s_0(t) = \left(\frac{t}{a}\right)^p x_0$ and $x_0 \in X$ is a limit point. If we denote

$$\phi(t) = |t^p| \left| \int_t^b u^{-p-1} \eta(u) du \right|,$$

then clearly $\phi : I \to [0, \infty)$ and

$$\|s(t) - s_0(t)\| \leq \phi(t). \tag{5.37}$$

Therefore, from (5.36) and (5.37) we get

$$\left\| t^3 f'''(t) + \alpha_1 t^2 f''(t) + \beta_1 t f'(t) + \gamma_1 f(t) - a^{-p} t^p x_0 \right\| \leq \phi(t). \tag{5.38}$$

Using Theorem 46 in (5.38), it follows that there exists a unique solution $f_0 : I \to X$ such that

$$\|f(t) - f_0(t)\| \leq |t^n| \left| \int_t^b u^{-n-1} \theta(u) du \right|,$$

where

$$\theta(t) = |t^l| \left| \int_t^b u^{-l-1} \psi(u) du \right|, \quad \psi(t) = |t^m| \left| \int_t^b u^{-m-1} \phi(u) du \right|,$$

$$\phi(t) = |t^p| \left| \int_t^b u^{-p-1} \eta(u) du \right|.$$

Here, $f_0(t)$ can be made any of the following cases
(i)

$$f_0(t) = \frac{e_1}{l-n} t^l + \frac{e_2}{m-n} t^m + \left\{ \frac{\bar{x}}{a^n} - \frac{e_1 a^{l-n}}{l-n} - \frac{e_2 a^{m-n}}{m-n} - \frac{e_3 a^{p-n}}{p-n} \right\} t^n + \frac{x_0}{a^p (p-l)(p-m)(p-n)} t^p,$$

where

$$e_1 = \left\{ \frac{x_1}{a^l} + \frac{x_0}{a^l (p-l)(m-l)} - \frac{x}{a^l (m-l)} \right\},$$

$$e_2 = \left\{ \frac{x}{a^m (m-l)} + \frac{x_0}{a^m (p-m)(l-m)} \right\} \quad \text{and} \quad e_3 = \frac{x_0}{a^p (p-l)(p-m)};$$

(ii)

$$f_0(t) = \left(\frac{e_1}{l-n} \right) t^l + \left\{ \frac{\bar{x}}{a^n} - \frac{e_1 a^{l-n}}{l-n} - \frac{e_3 a^{p-n}}{p-n} + e_2 ln \left| \frac{t}{a} \right| \right\} t^n + \frac{x_0}{a^p (p-l)(p-m)^2} t^p;$$

(iii)

$$f_0(t) = \left\{ \frac{e_4}{l-n} - \frac{e_5}{(l-n)^2} + \frac{e_5}{l-n} ln \left| \frac{t}{a} \right| \right\} t^l + \left\{ \frac{\bar{x}}{a^n} - \frac{e_4 a^{l-n}}{l-n} + \frac{e_5 a^{l-n}}{(l-n)^2} - \frac{e_6 a^{p-n}}{p-n} \right\} t^n + \frac{x_0}{a^p (p-n)(p-l)^2} t^p,$$

where

$$e_4 = \left\{ \frac{x_2}{a^l} - \frac{x_0}{a^l (p-l)} \right\}, e_5 = \left\{ \frac{x}{a^l} - \frac{x_0}{a^l (p-l)} \right\} \quad \text{and} \quad e_6 = \frac{x_0}{a^p (p-l)^2};$$

(iv)

$$f_0(t) = \left(\frac{e_2}{m-n} \right) t^m + \left\{ \frac{\bar{x}}{a^n} - \frac{e_2 \ a^{m-n}}{m-n} - \frac{e_3 \ a^{p-n}}{p-n} + e_1 ln \left| \frac{t}{a} \right| \right\} t^n + \frac{x_0}{a^p (p-m)(p-n)^2} t^p;$$

(v)

$$f_0(t) = \left\{ \frac{\bar{x}}{a^n} - \frac{e_6\ a^{p-n}}{p-n} + e_4\ ln\left|\frac{t}{a}\right| + \frac{e_5}{2}\left(ln\left|\frac{t}{a}\right|\right)^2 \right\} t^l + \frac{x_0}{a^p(p-l)^3}\ t^p;$$

(vi)

$$f_0(t) = \frac{e_7}{l-n}t^l + \frac{e_8}{m-n}t^m + \left\{ \frac{\bar{x}}{a^n} - \frac{e_7 a^{l-n}}{l-n} - \frac{e_8 a^{m-n}}{m-n} - \frac{e_9 a^{p-n}}{(p-n)^2} \right\} t^n + \left[\frac{e_9}{p-n} ln\left|\frac{t}{a}\right| - \frac{e_9}{(p-n)^2} \right] t^p,$$

where

$$e_7 = \left\{ \frac{x_3}{a^l} - \frac{x_0}{a^l(l-m)^2} + \frac{x}{a^l(l-m)} \right\},\quad e_8 = \left\{ \frac{x}{a^m(m-l)} + \frac{x_0}{a^m(m-l)^2} \right\}$$

and $e_9 = \dfrac{x_0}{a^p(p-m)}$;

(vii)

$$f_0(t) = \frac{e_1}{l-n}t^l + \frac{e_2}{m-n}t^m + \left\{ \frac{\bar{x}}{a^n} - \frac{e_1 a^{l-n}}{l-n} - \frac{e_2 a^{m-n}}{m-n} \right\} t^n + \frac{x_0}{a^p(p-l)(p-m)}\ t^p\ ln\left|\frac{t}{a}\right|;$$

(viii)

$$f_0(t) = \left(\frac{e_{10}}{l-n}\right) t^l + \left(\frac{e_{11}}{m-n}\right) t^m + \left\{ \frac{e_{12}}{p-n}\ ln\left|\frac{t}{a}\right| - \frac{e_{12}}{(p-n)^2} \right\} t^p + \left\{ \frac{\bar{x}}{a^n} - \frac{e_{10} a^{l-n}}{l-n} - \frac{e_{11} a^{m-n}}{m-n} + \frac{x_0}{a^n(m-l)(m-n)^2} \right\} t^n,$$

where

$$e_{10} = \frac{x_4}{a^l} - \frac{x}{a^l(m-l)} + \frac{x_0}{a^l(m-l)^2},\quad e_{11} = \frac{x}{a^m(m-l)} - \frac{x_0}{a^m(m-l)^2}$$

and $e_{12} = \dfrac{x_0}{a^p(p-l)}$;

(ix)

$$f_0(t) = \left(\frac{e_7}{l-n} - \frac{e_9}{p-n}\right) t^l + \left\{ \frac{\bar{x}}{a^n} - \frac{e_7\ a^{l-n}}{l-n} + \frac{e_9\ a^{p-n}}{(p-n)^2} + e_8 ln\left|\frac{t}{a}\right| \right\} t^n + \frac{x_0}{a^p(p-m)(p-n)}\ t^p\ ln\left|\frac{t}{a}\right|;$$

(x)

$$f_0(t) = \left\{ \frac{e_4}{l-n} - \frac{e_5}{(l-n)^2} + \frac{e_5}{l-n} ln\left|\frac{t}{a}\right| \right\} t^l + \left\{ \frac{\bar{x}}{a^n} - \frac{e_4\ a^{l-n}}{l-n} + \frac{e_5\ a^{l-n}}{(l-n)^2} \right\} t^n + \frac{x_0}{a^p(p-l)^2}\ t^p\ ln\left|\frac{t}{a}\right|;$$

(xi)

$$f_0(t) = \left\{ \frac{e_{11}}{m-n} - \frac{e_{12}}{(m-n)^2} \right\} t^m + \left\{ \frac{\bar{x}}{a^n} - \frac{e_{11}\ a^{m-n}}{m-n} + \frac{e_{12}\ a^{m-n}}{(m-n)^2} + e_{10}\ ln\left|\frac{t}{a}\right| \right\} t^n + \frac{x_0}{a^p(p-l)(p-n)}\ t^p\ ln\left|\frac{t}{a}\right|;$$

(xii)

$$f_0(t) = \left\{ \frac{x_5}{a^l(l-n)} - \frac{x}{a^l(l-n)^2} + \frac{x}{a^l(l-n)} ln\left|\frac{t}{a}\right| \right\} t^l + \left\{ \frac{\bar{x}}{a^n} - \frac{x_5}{a^n(l-n)} + \frac{x}{a^n(l-n)^2} - \frac{x_0}{a^n(l-n)^3} \right\} t^n + \left\{ \frac{x_0}{a^p(p-n)^3} - \frac{x_0}{a^p(p-n)^2} ln\left|\frac{t}{a}\right| + \frac{x_0}{2a^p(p-n)} \left(ln\left|\frac{t}{a}\right| \right)^2 \right\}\ t^p;$$

$(xiii)$

$$f_0(t) = \left(\frac{e_{10}}{l-n} \right) t^l + \left\{ \frac{\bar{x}}{a^n} - \frac{e_{10}\ a^{l-n}}{l-n} + e_{11}\ ln\left|\frac{t}{a}\right| \right\} t^n + \frac{x_0}{2a^p(p-l)} t^p \left(ln\left|\frac{t}{a}\right| \right)^2;$$

(xiv)

$$f_0(t) = \left(\frac{e_8}{m-n} \right) t^m + \left\{ \frac{\bar{x}}{a^n} - \frac{e_8\ a^{m-n}}{m-n} + e_7\ ln\left|\frac{t}{a}\right| \right\} t^n + \frac{x_0}{2a^p(p-m)} t^p \left(ln\left|\frac{t}{a}\right| \right)^2;$$

and
(xv)

$$f_0(t) = \left\{ \frac{\bar{x}}{a^l} + \frac{x}{a^l}\ ln\left|\frac{t}{a}\right| + \frac{x}{2a^l} \left(ln\left|\frac{t}{a}\right| \right)^2 + \frac{x_0}{6a^l} \left(ln\left|\frac{t}{a}\right| \right)^3 \right\} t^l,$$

where $x, \bar{x}, x_0, x_1, x_2, x_3, x_4$ and x_5 are limit points in X. These are the unique solutions of all possible cases. In fact, all these are the possible solutions of (5.27). Hence the theorem is proved.

Example 12 *Let $X = \mathbb{R}$ be a Banach space and $I = (a, \infty)$, $a > 0$. Consider the Euler's equation*

$$t^4 y^{(iv)}(t) + 2t^2 y^{''}(t) - 4ty^{'}(t) + 4y(t) = 0 \tag{5.39}$$

If we compare (5.39) with (5.27), then $\alpha = 0, \beta = 2, \gamma = -4$ and $\delta = 4$, and $l = 1, m = 1, n = 2, p = 2$ are the characteristic roots of (5.39). Let $f : I \to X$ be satisfy the differential inequality

$$\left\| t^4 f^{(iv)}(t) + 2f^2 y^{''}(t) - 4tf^{'}(t) + 4f(t) \right\| \le \epsilon$$

for any $\epsilon > 0$ and for any $t \in I$. Then, by Theorem 47, there exists a unique solution $f_0 : I \to X$ such that

$$\|f(t) - f_0(t)\| \le |t^2| \left| \int_t^b u^{-3}\ \theta(u) du \right|,$$

where $I = (a, b)$ and

$$\phi(t) = |t^2| \left| \int_t^b u^{-3}\ \epsilon du \right| = \frac{\epsilon}{2} \left| \left(\frac{t}{b} \right)^2 - 1 \right|.$$

When $b \to \infty, \phi(t) = \frac{\epsilon}{2}$. Also

$$\psi(t) = |t| \left| \int_t^b u^{-2}\ \frac{\epsilon}{2} du \right| = \frac{\epsilon}{2} \left| \left(\frac{t}{b} \right) - 1 \right| = \frac{\epsilon}{2}$$

and

$$\theta(t) = |t| \left| \int_t^b u^{-2}\ \frac{\epsilon}{2} du \right| = \frac{\epsilon}{2}$$

as $b \to \infty$. Hence,

$$\|f(t) - f_0(t)\| \le |t^2| \left| \int_t^b u^{-3}\ \frac{\epsilon}{2} du \right| = \frac{\epsilon}{4} \left| \left(\frac{t}{b} \right)^2 - 1 \right|.$$

When $b \to \infty$,

$$\|f(t) - f_0(t)\| \le \frac{\epsilon}{4}$$

in which

$$f_0(t) = \left[\left(\frac{2x_0 - x - x_2}{a} \right) + \left(\frac{x_0 - x_2}{a} \right) ln \left| \frac{t}{a} \right| \right] t + \left[\left(\frac{\bar{x} + x + x_2 - 2x_0}{a^2} \right) + \left(\frac{x_0}{a^2} \right) ln \left| \frac{t}{a} \right| \right] t^2$$

and $x, \bar{x}, x_0, x_2$ are the unique elements of X.

5.5 Notes

The results of Sections 5.2 and 5.3 for Hyers-Ulam stability of first and second order Euler's differential equations are extracted from the work of S. M. Jung [18] except Theorem 42. It is noticed that Theorem 42 plays an important role to go for the Hyers-Ulam stability of third and fourth order Euler's differential equations and these results are depicted from the works of A. K. Tripathy and A. Satapathy [58] and [59]. It would be interesting to extend the works of Tripathy and Satapathy to fifth order and sixth order Euler's differential equations of the form:

$$t^5y^{(v)} + \alpha t^4y^{(iv)} + \beta t^3y^{'''} + \gamma t^2y^{''} + \eta ty' + \sigma y = 0$$

$$t^6y^{(vi)} + \alpha t^5y^{(v)} + \beta t^4y^{iv} + \gamma t^3y^{'''} + \eta t^2y'' + \sigma ty' + \tau y = 0.$$

Chapter 6

Generalized Hyers-Ulam Stability of Differential Equations in Complex Banach Space

This chapter deals with the Hyers-Ulam stability of linear first and second order complex differential equations

$$f'(z) + p(z)f(z) + q(z) = 0, \tag{6.1}$$

$$y''(t) + \alpha y'(t) + \beta y(t) = 0 \tag{6.2}$$

and

$$y''(t) + \alpha y'(t) + \beta y(t) = \sigma(x) \tag{6.3}$$

in Banach space, where $y \in C^2(I)$, $\alpha, \beta \in \mathbb{C}$, $f \in C^1(I), \sigma \in C(I)$, $I = (a, b)$, $-\infty < a < b < +\infty$.

6.1 Hyers-Ulam Stability of First Order Differential Equations

In this section, we present the **Hyers-Ulam** stability of the Banach space-valued differential equations of the form 6.1. Let X be a complex Banach space and let Ω be an open set of $\mathbb{C}$. A mapping $f : \Omega \longrightarrow X$ is said to be **holomorphic** if and only if

$$\lim_{w \longrightarrow z} \frac{f(w) - f(z)}{w - z}$$

exists in the norm-topology of X for each $z \in \Omega$. We know that $f : \Omega \longrightarrow X$ is holomorphic if and only if $\phi\ of : \Omega \longrightarrow \mathbb{C}$ is holomorphic (as a complex-valued function) for each $\phi \in X^*$, the dual space of X. We denote $H(\Omega, X)$, the set of all holomorphic mappings $f : \Omega \longrightarrow X$. In short, $H(\Omega) = H(\Omega, \mathbb{C})$. Let $f \in H(\Omega, X)$. Let's consider the case where $0 \in \Omega$. For $z \in \Omega$, we write $\int_0^z f(\zeta)d\zeta$ for $\int_0^1 zf(zt)dt$, the integral of f over the path γ defined by

$$\gamma(t) = zt, \ \ \forall \in [0, 1].$$

For each $p \in H(\Omega)$, define $\tilde{p}$ as

$$\tilde{p}(z) = exp \int_0^z p(\zeta)d\zeta, \ \ \forall z \in \Omega.$$

If Ω is convex, then $\tilde{p} \in H(\Omega)$ and

$$\tilde{p}'(z) = p(z)\tilde{p}(z), \ \ \forall z \in \Omega.$$

Lemma 12 *Let Ω be a convex open set of $\mathbb{C}$ with $0 \in \Omega$. Let X be a complex Banach space and $p \in H(\Omega)$. For $f, q, u \in H(\Omega, X)$, each of the following conditions are equivalent.*

(a) $f'(z) + p(z)f(z) + q(z) = u(z)$, *for all* $z \in \Omega$.

(b) $f(z) = \frac{1}{\tilde{p}(z)}\{f(0) + \int_0^z \tilde{p}(\zeta)(u(\zeta) - q(\zeta))d\zeta\}$, *for all* $z \in \Omega$.

Proof 80 *Consider the case: (a) implies (b). Let $f'(z) + p(z)f(z) + q(z) - u(z) = 0$ for every $z \in \Omega$. Since,*

$$\tilde{p}'(z) = p(z)\tilde{p}(z), \ \ z \in \Omega,$$

then,

$$\begin{aligned}
\{\tilde{p}(z)f(z)\}' &= \tilde{p}'(z)f(z) + \tilde{p}(z)f'(z) \\
&= p(z)\tilde{p}(z)f(z) + \tilde{p}(z)f'(z) \\
&= \tilde{p}(z)\{p(z)f(z) + f'(z)\} \\
&= \tilde{p}(z)(u(z) - q(z)), \ \ \forall z \in \Omega.
\end{aligned}$$

Integrating the above relation from 0 to z, we obtain

$$\int_0^z \{f(\zeta)\tilde{p}(\zeta)\}'dz = \int_0^z \tilde{p}(\zeta)(u(\zeta) - q(\zeta))d\zeta,$$

that is,

$$f(z)\tilde{p}(z) - f(0)\tilde{p}(0) = \int_0^z \tilde{p}(\zeta)(u(\zeta) - q(\zeta))d\zeta, \ \ \forall z \in \Omega.$$

By definition, $\tilde{p}(0) = 1$ and hence

$$f(z) = \frac{1}{\tilde{p}(z)}\left\{f(0) + \int_0^z \tilde{p}(\zeta)(u(\zeta) - q(\zeta))d\zeta\right\}, \ \ \forall z \in \Omega.$$

Next, we consider the case for (b) implies (a). Let

$$f(z) = \frac{1}{\tilde{p}(z)}\left\{f(0) + \int_0^z \tilde{p}(\zeta)(u(\zeta) - q(\zeta))d\zeta\right\}, \ \ \forall z \in \Omega.$$

Then

$$f'(z) = \frac{1}{\tilde{p}^2(z)}\left[\left\{f(0) + \int_0^z \tilde{p}(\zeta)(u(\zeta) - q(\zeta))d\zeta\right\}'\tilde{p}(z) - \left\{f(0) + \int_0^z \tilde{p}(\zeta)(u(\zeta) - q(\zeta))d\zeta\right\}\tilde{p}'(z)\right], \quad \forall z \in \Omega.$$

Hence,

$$f'(z) = \frac{1}{\tilde{p}^2(z)}\{\tilde{p}^2(z)(u(z) - q(z)) - \tilde{p}(z)f(z)\tilde{p}'(z)\},$$

that is,

$$f'(z) + p(z)f(z) + q(z) = u(z), \quad \forall z \in \Omega.$$

Hence, the lemma is proved.

Theorem 48 *Let Ω be a convex open set of $\mathbb{C}$ with $0 \in \Omega$. Let X be a complex Banach space and $q \in H(\Omega, X)$. Suppose that $p \in H(\Omega)$ satisfies that*

$$C_p = \sup_{z\in\Omega} \frac{1}{\tilde{p}(z)}\left|\int_0^z |\tilde{p}(\zeta)|d\zeta\right| < \infty.$$

For each $\varepsilon \geq 0$ and $f \in H(\Omega, X)$ satisfying

$$||f'(z) + p(z)f(z) + q(z)|| \leq \varepsilon, \quad \forall z \in \Omega \tag{6.4}$$

there exists $g \in H(\Omega, X)$ such that

$$g'(z) + p(z)g(z) + q(z) = 0, \quad \forall z \in \Omega$$

and that

$$||f(z) - g(z)|| \leq C_p\varepsilon, \quad \forall z \in \Omega.$$

Proof 81 *Suppose that $\varepsilon \geq 0$ and $f \in H(\Omega, X)$, such that*

$$||f'(z) + p(z)f(z) + q(z)|| \leq \varepsilon, \quad \forall z \in \Omega.$$

If $u(z) = f'(z) + p(z)f(z) + q(z), \ \forall z \in \Omega$, then by Lemma 12,

$$f(z) = \frac{1}{\tilde{p}(z)}\left\{f(0) + \int_0^z \tilde{p}(\zeta)(u(\zeta) - q(\zeta))d\zeta\right\}, \quad \forall z \in \Omega. \tag{6.5}$$

For all $z \in \Omega$, define

$$g(z) = \frac{1}{\tilde{p}(z)}\left\{f(0) - \int_0^z \{\tilde{p}(\zeta)q(\zeta))d\zeta\right\}.$$

Then $g \in H(\Omega, X)$ satisfies

$$g'(z) + p(z)g(z) + q(z) = 0, \quad \forall z \in \Omega$$

due to Lemma 12. Now,

$$\begin{aligned}||f(z)-g(z)|| &\le \frac{1}{|\tilde{p}(z)|}\left|\left|\int_0^z \tilde{p}(\zeta)u(\zeta)d\zeta\right|\right| \\ &\le \frac{\varepsilon}{|\tilde{p}(z)|}\left|\int_0^z |\tilde{p}(\zeta)|d\zeta\right|, \ \forall z \in \Omega,\end{aligned}$$

because $||u(z)|| \le \varepsilon, \ \forall z \in \Omega$. *Hence,*

$$||f(z)-g(z)|| \le \varepsilon \sup_{z\in\Omega} \frac{1}{|\tilde{p}(z)|}\left|\int_0^z |\tilde{p}(\zeta)|d\zeta\right| = C_p\varepsilon, \ \forall z \in \Omega.$$

Thus, the proof is complete.

Theorem 49 *Let* Ω *be a convex open set of* $\mathbb{C}$ *with* $0 \in \Omega$. *Let* X *be a complex Banach space and* $q \in H(\Omega, X)$ *and* $p \in H(\Omega)$. *Suppose that there exists* $\lambda \in \partial\Omega$, *the boundary of* Ω *such that*

$$D_p(\lambda) = \sup_{z\in\Omega} \frac{1}{|\tilde{p}(z)|}\left|\int_\lambda^z |\tilde{p}(\zeta)|d\zeta\right| < \infty,$$

where

$$\int_\lambda^z |\tilde{p}(\zeta)|d\zeta = \lim_{a\to 0}\int_a^1 |\tilde{p}(\lambda + t(z-\lambda))|(z-\lambda)dt.$$

For each $\varepsilon \ge 0$ *and* $f \in H(\Omega, X)$ *satisfying* (6.4), *there exists* $g \in H(\Omega, X)$ *such that*

$$g'(z) + p(z)g(z) + q(z) = 0, \ \ (\forall z \in \Omega)$$

and that

$$\|f(z) - g(z)\| \le D_p(\lambda)\varepsilon, \ \forall z \in \Omega.$$

Proof 82 *Let there exist* $\lambda \in \partial\Omega$ *such that*

$$D_p(\lambda) = \sup_{z\in\Omega} \frac{1}{|\tilde{p}(z)|}\left|\int_\lambda^z |\tilde{p}(\zeta)|d\zeta\right| < \infty.$$

For $\varepsilon \ge 0$ *and* $f \in H(\Omega, X)$, *let*

$$||f'(z) + p(z)f(z) + q(z)|| \le \varepsilon, \ \ (\forall z \in \Omega).$$

For all $z \in \Omega$, *let's set*

$$v(z) = f'(z) + p(z)f(z) + q(z).$$

Then $\|v(z)\| \le \varepsilon, \ \forall z \in \Omega$. *By Lemma 12,*

$$f(z) = \frac{1}{\tilde{p}(z)}\left\{f(0) + \int_0^z \{\tilde{p}(\zeta)(v(\zeta) - q(\zeta))d\zeta\right\}, \ \forall z \in \Omega. \tag{6.6}$$

But Ω is convex, so is $\bar{\Omega}$ and the closure of Ω. Hence for $0 < t \leq 1$,

$$\lambda + t(z-\lambda) \in \Omega, \ \ \forall z \in \Omega.$$

By the hypothesis, $\tilde{p}v \in H(\Omega, X)$ and by using Cauchy theorem

$$\int_0^z \tilde{p}(\zeta)v(\zeta)d\zeta = \int_0^\lambda \tilde{p}(\zeta)v(\zeta)d\zeta + \int_\lambda^z \tilde{p}(\zeta)v(\zeta)d\zeta, \ \ \forall z \in \Omega. \tag{6.7}$$

Setting

$$g(z) = \frac{1}{\tilde{p}(z)}\left\{f(0) + \int_0^\lambda \tilde{p}(\zeta)v(\zeta)d\zeta - \int_0^z \tilde{p}(\zeta)q(\zeta)d\zeta\right\}, \ \ \forall z \in \Omega, \tag{6.8}$$

it follows that

$$g'(z) + p(z)g(z) + q(z) = 0, \ \ (\forall z \in \Omega).$$

By (6.6),(6.7) and (6.8), we obtain

$$\begin{aligned}\|f(z) - g(z)\| &\leq \frac{1}{|\tilde{p}(z)|}\left\|\int_\lambda^z \tilde{p}(\zeta)v(\zeta)d\zeta\right\| \\ &\leq \varepsilon \sup_{z\in\Omega} \frac{1}{|\tilde{p}(z)|}\left|\int_\lambda^z \tilde{p}(\zeta)d\zeta\right| = D_p(\lambda)\varepsilon, \ \ \forall z \in \Omega.\end{aligned}$$

This completes the proof of the theorem.

Example 13 *Let Ω be a bounded convex open set of $\mathbb{C}$, say $|z| \leq M$ for all $z \in \Omega$. If $0 \in \Omega$ and $p \in H(\Omega)$ is bounded, then*

$$C_p = \sup_{z\in\Omega} \ \frac{1}{|\tilde{p}(z)|}\left|\int_0^z |\tilde{p}(\zeta)|d\zeta\right| < \infty.$$

If $p \in H(\Omega)$ is bounded such that $|p(z)| \leq K$ for all $z \in \Omega$, then

$$\left|\int_0^z p(\zeta)d\zeta\right| \leq \int_0^1 |p(\zeta t)||z|dt \leq KM, \forall z \in \Omega.$$

Hence,

$$|\tilde{p}(\zeta)| = exp\left(Re\int_0^z p(\zeta)d\zeta\right) \geq e^{-KM}, \ \ \forall z \in \Omega. \tag{6.9}$$

As $|\tilde{p}(\zeta)| \leq e^{KM}, \ \ \forall z \in \Omega$, then we obtain

$$\left|\int_0^z |\tilde{p}(z)|d\zeta\right| \leq Me^{KM}, \ \ \forall z \in \Omega. \tag{6.10}$$

It follows from (6.9) and (6.10) that

$$C_p = \sup_{z\in\Omega} \ \frac{1}{|\tilde{p}(z)|}\left|\int_0^z |\tilde{p}(\zeta)|d\zeta\right| \leq Me^{2KM} < \infty.$$

Example 14 *Let $\Omega = \{z \in \mathbb{C} : |z| < 1\}$. Let $p \in H(\Omega)$ be defined by, $p(z) = \frac{1}{z+1}$, $\forall z \in \Omega$. Then $log(z+1)$, $\forall z \in \Omega$ is well-defined such that*

$$\tilde{p}(z) = exp \int_0^z p(\zeta)d\zeta = e^{log(z+1)} = z+1, \ \forall z \in \Omega. \tag{6.11}$$

Hence,

$$\begin{aligned} \left| \int_{-1}^z |\tilde{p}(\zeta)|d\zeta \right| &= \lim_{a\to 0} \left| \int_a^1 |\tilde{p}(-1+t(z+1))|(z+1)dt \right| \\ &= \lim_{a\to 0} \left| \int_a^1 |t(z+1)|(z+1)dt \right| \\ &= |z+1|^2 \int_0^1 tdt = \frac{|z+1|^2}{2}, \ \forall z \in \Omega. \end{aligned} \tag{6.12}$$

By (6.11) and (6.12), we conclude that

$$D_p(-1) = sup_{z\in\omega} \frac{1}{|\tilde{p}|} \left| \int_{-1}^z |\tilde{p}(\zeta)|d\zeta \right| = sup_{z\in\Omega} \frac{|z+1|}{2} = 1.$$

We note that

$$D_p(\lambda) = sup_{z\in\omega} \frac{1}{|\tilde{p}|} \left| \int_{\lambda}^z |\tilde{p}(\zeta)|d\zeta \right| = \infty, \ \forall \lambda \in \partial\Omega \backslash \{-1\}. \tag{6.13}$$

Therefore,

$$C_p = sup_{z\in\omega} \frac{1}{|\tilde{p}|} \left| \int_0^z |\tilde{p}(\zeta)|d\zeta \right| = \infty. \tag{6.14}$$

Consider that $\lambda \in \partial\Omega \backslash \{-1\}$. For all $n \in \mathbb{N}$, we can associate corresponding $z_n \in \Omega$ such that

$$|\lambda + 1| \geq |\lambda - z_n| \ and \ |z_n + 1| < \frac{1}{n}. \tag{6.15}$$

Hence, for all $n \in \mathbb{N}$

$$\begin{aligned} \left| \int_{\lambda}^{z_n} |\tilde{p}(\zeta)|d\zeta \right| &= \lim_{a\to 0} \left| \int_a^1 |\tilde{p}(\lambda + t(z_n - \lambda))|(z_n - \lambda)dt \right| \\ &= \lim_{a\to 0} \left| \int_a^1 |\lambda + 1 + t(z_n - \lambda)|(z_n - \lambda)dt \right| \\ &\geq |z_n - \lambda| \int_0^1 ||\lambda + 1| - t|z_n - \lambda||dt. \end{aligned}$$

By (6.15), we get

$$\begin{aligned} \int_0^1 ||\lambda + 1| - t|z_n - \lambda||dt &= \int_0^1 ||\lambda + 1| - t|z_n - \lambda||dt \\ &= |\lambda + 1| - \frac{z_n - \lambda}{2} \end{aligned}$$

and

$$\left|\int_{\lambda}^{z_n} |\tilde{p}(\zeta)|d\zeta\right| \geq |z_n - \lambda|\left(|\lambda + 1| - \frac{z_n - \lambda}{2}\right) \tag{6.16}$$

On the other hand, (6.15) *and* (6.16) *give rise to*

$$\left|\int_{\lambda}^{z_n} |\tilde{p}(\zeta)|d\zeta\right| \geq |z_n - \lambda|\frac{|z_n - \lambda|}{2} = \frac{|z_n - \lambda|^2}{2}.$$

Consequently,

$$\frac{1}{|\tilde{p}(z_n)|}\left|\int_{\lambda}^{z_n} |\tilde{p}(\zeta)|d\zeta\right| \geq \frac{|z_n - \lambda|^2}{2|z_n + 1|}$$

$$\rightarrow \infty, \text{ as } n \rightarrow \infty$$

due to $z_n \rightarrow -1$ *for* $n \rightarrow \infty$. *Therefore,* (6.13) *is true for all* $\lambda \in \partial\Omega\backslash\{-1\}$. *Because* $|z| < 1$ *for all* $z \in \Omega$, *it happens that*

$$\begin{aligned}\left|\int_{0}^{z} |\tilde{p}(\zeta)|d\zeta\right| &= |z|\int_{0}^{1} |\tilde{p}(tz)|dt = |z|\int_{0}^{1} |tz + 1|dt \\ &\geq |z|\int_{0}^{1} ||tz| - 1|dt = |z|\int_{0}^{1} |(1 - |z|t)dt \\ &= |z|\left(1 - \frac{|z|}{2}\right).\end{aligned} \tag{6.17}$$

From (6.11) *and* (6.17), *we conclude that*

$$C_p = \sup_{z \in \Omega} \frac{1}{|\tilde{p}(z)|}\left|\int_{0}^{z} |\tilde{p}(\zeta)|d\zeta\right| = sup_{z \in \Omega}\frac{|z|(2 - |z|)}{2|z + 1|} = \infty.$$

6.2 Hyers-Ulam Stability of Second Order Differential Equations

Lemma 13 *Let* X *be a complex Banach space and let* $I = (a, b)$ *be an open interval, where* $a, b \in \mathbb{R}$ *are arbitrarily given with* $-\infty < a < b < +\infty$. *Assume that* g *is an arbitrarily complex function,* $h : I \rightarrow \mathbf{C}$ *is continuous and integrable on* I. *Moreover, suppose* $\varphi : I \rightarrow [0, \infty)$ *is an integrable function on* I. *If a continuously differentiable function* $y : I \rightarrow X$ *satisfies the differential inequality*

$$\|y'(t) + gy(t) + h(t)\| \leq \varphi(t), \tag{6.18}$$

for all $t \in I$, *then there exists a unique* $x \in X$ *such that*

$$\left\|y(t) - e^{-\int_a^t gdu}\left(x - \int_{a}^{t} e^{\int_a^v gdu}h(v)dv\right)\right\| \leq e^{-\Re\int_a^t gdu}\int_{t}^{b} \varphi(v)e^{-\Re\int_a^v gdu}dv.$$

Proof 83 *Consider,*

$$z(t) = e^{\int_a^t g du} y(t) + \int_a^t e^{\int_a^v g du} h(v) dv,$$

for all $t \in I$. *Now,*

$$\begin{aligned}
\|z(t) - z(s)\| &= \left\| e^{\int_a^t g du} y(t) + \int_a^t e^{\int_a^v g du} h(v) dv \right. \\
&\quad \left. - e^{\int_a^s g du} y(s) - \int_a^s e^{\int_a^v g du} h(v) dv \right\| \\
&= \left\| e^{\int_a^t g du} y(t) - e^{\int_a^s g du} y(s) + \int_s^t e^{\int_a^v} h(v) dv \right\| \\
&= \left\| \int_s^t \frac{d}{dv} [e^{\int_a^v g du} y(v)] dv + \int_s^t e^{\int_a^v} h(v) dv \right\| \\
&= \left\| \int_s^t [e^{\int_a^v g du} y(v) g + y'(v) e^{\int_a^v g du}] dv + \int_s^t e^{\int_a^v} h(v) dv \right\| \\
&= \left\| \int_s^t e^{\int_a^v g du} [y'(v) + g y(v) + h(v)] dv \right\| \\
&\leq \left| \int_s^t e^{\Re \int_a^v g du} \varphi(v) dv \right|,
\end{aligned}$$

for any $s, t \in I$. *Since* g *is an arbitrary complex number and* $\varphi(v)$ *is integrable, then* $e^{\Re \int_a^v g du} \varphi(v)$ *is integrable on* I. *Hence there exists an* $x \in X$ *such that* $z(s)$ *converges to* x *as* $s \to b$.

$$\begin{aligned}
&\left\| y(t) - e^{-\int_a^t g du} \left(x - \int_a^t e^{\int_a^v g du} h(v) dv \right) \right\| \\
&= \| e^{-\int_a^t g du} [z(t) - x] \| \\
&\leq e^{-\Re \int_a^t g du} \| z(t) - z(s) \| + e^{-\Re \int_a^t g du} \| z(s) - x \| \\
&\leq e^{-\Re \int_a^t g du} \left| \int_s^t e^{\Re \int_a^v g du} \varphi(v) dv \right| + e^{-\Re \int_a^t g du} \| z(s) - x \| \\
&= e^{-\Re \int_a^t g du} \left| \int_t^b e^{\Re \int_a^v g du} \varphi(v) dv \right|,
\end{aligned}$$

where $z(s) \to x$ *as* $s \to b$, *then*

$$\begin{aligned}
&\left\| y(t) - e^{-\int_a^t g du} \left(x - \int_a^t e^{\int_a^v g du} h(v) dv \right) \right\| \\
&\qquad \leq e^{-\Re \int_a^t g du} \int_t^b e^{\Re \int_a^v g du} \varphi(v) dv, \qquad (6.19)
\end{aligned}$$

for all $t \in I$. *If* $y_0(t) = e^{-\int_a^t g du} \left(x - \int_a^t e^{\int_a^v g du} h(v) dv \right)$, *then*

$y_0'(t) = -xge^{-\int_a^t gdu} + ge^{-\int_a^t gdu}\int_a^t e^{\int_a^v gdu}h(v)dv - h(t)$ *and*

$$y_0'(t) + gy_0(t) + h(t) = -xge^{-\int_a^t gdu} + ge^{-\int_a^t gdu}\int_a^t e^{\int_a^v gdu}h(v)dv - h(t)$$
$$+ xge^{-\int_a^t gdu} - ge^{-\int_a^t gdu}\int_a^t e^{\int_a^v gdu}h(v)dv + h(t) = 0.$$

So $y_0(t)$ *satisfying* $y_0'(t) + gy_0(t) + h(t) = 0$. *It remains to show that* x *is unique. Let* $x_1 \in X$. *Then*

$$\begin{aligned}\|e^{-\int_a^t gdu}(x_1 - x)\| &= \left\|y(t) - e^{-\int_a^t gdu}\left(x - \int_a^t e^{\int_a^v gdu}h(v)dv\right)\right.\\ &\quad \left. - y(t) + e^{-\int_a^t gdu}\left(x_1 - \int_a^t e^{\int_a^v gdu}h(v)dv\right)\right\|\\ &\leq \left\|y(t) - e^{-\int_a^t gdu}\left(x - \int_a^t e^{\int_a^v gdu}h(v)dv\right)\right\|\\ &\quad + \left\|-\left\{y(t) - e^{-\int_a^t gdu}\left(x - \int_a^t e^{\int_a^v gdu}h(v)dv\right)\right\}\right\|\\ &\leq 2e^{-\Re\int_a^t gdu}\left|\int_t^b e^{\Re\int_a^v gdu}\varphi(v)dv\right|\end{aligned}$$

implies that

$$\|x_1 - x\| \leq \frac{2e^{-\Re\int_a^t gdu}\left|\int_t^b e^{\Re\int_a^v gdu}\varphi(v)dv\right|}{e^{-\Re\int_a^t gdu}},$$

that is,

$$\|x_1 - x\| \leq 2\left|\int_t^b e^{\Re\int_a^v gdu}\varphi(v)dv\right| \to 0$$

as $t \to b$. *Hence* $x_1 = x$.

Corollary 27 *Let* X *be a complex Banach space and let* $I = (a, b)$ *be an open interval, where* $a, b \in \mathbb{R}$ *are arbitrarily given with* $-\infty < a < b < +\infty$. *Assume that* g *is an arbitrarily complex number,* $h : I \to \mathbf{C}$ *is continuous and integrable on* I. *Moreover, suppose* $\varphi : I \to [0, \infty)$ *is an integrable function on* I. *If a continuously differentiable function* $y : I \to X$ *satisfies the differential inequality*

$$\|y'(t) + gy(t) + h(t)\| \leq \varphi(t), \tag{6.20}$$

for all $t \in I$, *then there exists a unique* $x \in X$ *such that*

$$\left\|y(t) - e^{-\int_b^t gdu}\left(x - \int_b^t e^{\int_b^v gdu}h(v)dv\right)\right\|$$
$$\leq e^{-\Re\int_b^t gdu}\int_a^t \varphi(v)e^{-\Re\int_a^v gdu}dv.$$

Proof 84 *Let $J = (-b, -a)$ and define $h_1(t) = h(-t)$, $y_1(t) = y(-t)$ and $\varphi_1(t) = \varphi(-t)$. Using these definition (6.20) becomes*

$$\|y_1'(t) - gy_1(t) - h_1(t)\| \leq \varphi_1(t),$$

for $t \in J$. Hence by using Lemma 13, there exists a unique $x \in X$ such that

$$\left\| y_1(t) - e^{\int_{-b}^{t} gdu}\left(x + \int_b^t e^{-\int_{-b}^{v} gdu} h_1(v)dv\right)\right\| \leq e^{\Re \int_{-b}^{t} gdu} \int_t^{-a} \varphi_1(v) e^{-\Re \int_{-b}^{v} gdu} dv,$$

that is,

$$\left\| y(t) - e^{-\int_b^t gdu}\left(x + \int_b^t e^{\int_b^v gdu} h(v)dv\right)\right\| \leq e^{-\Re \int_b^t gdu} \int_a^t \varphi(v) e^{\Re \int_b^v gdu} dv,$$

for any $t \in J$.

Theorem 50 *Let $\varphi : I \to [0, \infty)$ be an integrable function on I. Assume that α,β are complex numbers. If a twice continuously differentiable function $y(t)$ satisfies the inequality*

$$\|y''(t) + \alpha y'(t) + \beta y(t)\| \leq \varphi(t). \tag{6.21}$$

Then (6.2) has the Hyers-Ulam-Rassias stability.

Proof 85 *Let λ_1 and λ_2 be the roots of characteristic equation $\lambda^2+\alpha\lambda+\beta = 0$. Define $g(t) = y'(t) - \lambda_1 y(t)$, then $g'(t) = y''(t) - \lambda_1 y'(t)$. Now,*

$$\begin{aligned}\|g'(t) - \lambda_2 g(t)\| &= \|y''(t) - \lambda_1 y'(t) - \lambda_2\{y'(t) - \lambda_1 y(t)\}\| \\ &= \|y''(t) - \lambda_1 y'(t) - \lambda_2 y'(t) + \lambda_1\lambda_2 y(t)\| \\ &= \|y''(t) + (-\lambda_1 - \lambda_2)y'(t) + \lambda_1\lambda_2 y(t)\| \\ &= \|y'' + \alpha y'(t) + \beta y(t)\| \leq \varphi(t)\end{aligned}$$

implies that

$$\|g'(t) - \lambda_2 g(t)\| \leq \varphi(t).$$

By using Lemma 13, there exists a unique $x_1 \in X$ such that

$$\|g(t) - e^{\int_a^t \lambda_2 du}(x_1 - 0)\| \leq e^{\Re\Re \int_a^t \lambda_2 du} \int_t^b e^{-\Re \int_a^v \lambda_2 du} \varphi(v)dv$$

implies that

$$\|g(t) - x_1 e^{\lambda_2(t-a)}\| \leq e^{\Re\lambda_2(t-a)} \int_t^b e^{-\Re \int_a^v \lambda_2 du} \varphi(v)dv,$$

that is,

$$\|g(t) - x_1 e^{t\lambda_2 - a\lambda_2}\| \leq e^{\Re(t\lambda_2 - a\lambda_2)} \int_t^b e^{-\Re\int_a^v \lambda_2 du} \varphi(v) dv,$$

where $x_1 = \lim_{t\to b} g(t) e^{-\lambda_2 t + \lambda_2 a}$. *If* $u(t) = x_1 e^{t\lambda_2 - a\lambda_2}$, *then* $u'(t) = \lambda_2 x_1 e^{t\lambda_2 - a\lambda_2}$ *and*

$$u'(t) - \lambda_2 u(t) = 0. \tag{6.22}$$

Since $g(t) = y'(t) - \lambda_1 y(t)$, *then*

$$\|y'(t) - \lambda_1 y(t) - x_1 e^{t\lambda_2 - a\lambda_2}\| \leq e^{t\lambda_2 - a\lambda_2} \int_t^b e^{-\Re\int_a^v \lambda_2 du} \varphi(v) dv. \tag{6.23}$$

Let us define $\psi(t) = e^{\Re(t\lambda_2 - a\lambda_2)} \int_t^b e^{-\Re\int_a^v \lambda_2 du} \varphi(v) dv$ *and* $u(t) = x_1 e^{t\lambda_2 - a\lambda_2}$, *then* (6.23) *becomes*

$$\|y'(t) - \lambda_1 y(t) - u(t)\| \leq \psi(t).$$

Again, by using Lemma 13, there exists a unique $x_2 \in X$ *such that*

$$\left\| y(t) - e^{\int_a^t \lambda_1 du} \left(x_2 + \int_a^t e^{-\int_a^v \lambda_1 du} u(v) dv \right) \right\| \leq e^{\Re\int_a^t \lambda_1 du} \int_t^b \psi(v) e^{\Re\int_a^v \lambda_1 du},$$

where $x_2 = \lim_{t\to b}(e^{-\int_a^t \lambda_1 du} y(t) - \int_a^t e^{-\int_a^v \lambda_1 du} . x_1 . e^{v\lambda_2 - a\lambda_2} dv)$.
If $y_0(t) = e^{\int_a^t \lambda_1 du}(x_2 + \int_a^t e^{-\int_a^v \lambda_1 du} u(v) dv)$, *then*
$y_0'(t) = x_2 \lambda_1 e^{\int_a^t \lambda_1 du} + \lambda_1 e^{\int_a^t \lambda_1 du} \int_a^t e^{-\int_a^v \lambda_1 du} u(v) dv + u(t)$ *and*

$$y_0'(t) - \lambda_1 y_0(t) - u(t) = x_2 \lambda_1 e^{\int_a^t \lambda_1 du} + \lambda_1 e^{\int_a^t \lambda_1 du} \int_a^t e^{-\int_a^v \lambda_1 du} u(v) dv$$
$$+ u(t) - x_2 \lambda_1 e^{\int_a^t \lambda_1 du} + \lambda_1 e^{\int_a^t \lambda_1 du} \int_a^t e^{-\int_a^v \lambda_1 du} u(v) dv - u(t) = 0.$$

So $u(t) = y_0'(t) - \lambda_1 y_0(t)$ *and* $u'(t) = y_0''(t) - \lambda_1 y_0'(t)$. *Using this in* (6.22), *we get*

$$y_0''(t) - \lambda_1 y_0'(t) - \lambda_2\{y_0'(t) - \lambda_1 y_0(t)\} = 0,$$

implies that

$$y_0''(t) + (-\lambda_1 - \lambda_2) y_0'(t) + \lambda_1 \lambda_2 y_0(t) = 0,$$

that is,

$$y_0''(t) + \alpha y_0'(t) + \beta y_0(t) = 0.$$

This competes the proof.

Theorem 51 *Let $\varphi : I \to [0,\infty)$ be an integrable function on I. Assume that α,β are complex numbers, and $f : I \to X$ is continuous and integrable on I. If a twice continuously function $y(t)$ satisfies the inequality*

$$\|y''(t) + \alpha y'(t) + \beta y(t) - \sigma(t)\| \leq \varphi(t).$$

Then (6.3) has the Hyers-Ulam-Rassias stability.

Proof 86 *Let λ_1 and λ_2 be the roots of characteristic equation $\lambda^2+\alpha\lambda+\beta = 0$. Define $g(t) = y'(t) - \lambda_1 y(t)$. Then $g'(t) = y''(t) - \lambda_1 y'(t)$. Now,*

$$\begin{aligned}\|g'(t) - \lambda_2 g(t) - f(t)\| &= \|y''(t) - \lambda_1 y'(t) - \lambda_2(y'(t) - \lambda_1 y(t)) - \sigma(t)\| \\ &= \|y''(t) - \lambda_1 y'(t) - \lambda_2 y'(t) + \lambda_1\lambda_2 y(t) - \sigma(t)\| \\ &= \|y''(t) + (-\lambda_1 - \lambda_2)y'(t) + \lambda_1\lambda_2 y(t) - \sigma(t)\| \\ &= \|y'' + \alpha y'(t) + \beta y(t) - \sigma(t)\| \leq \varphi(t)\end{aligned}$$

implies that

$$\|g'(t) - \lambda_2 g(t) - \sigma(t)\| \leq \varphi(t).$$

By using Lemma 13, there exists a unique $x_1 \in X$ such that

$$\begin{aligned}&\left\| g(t) - e^{\int_a^t \lambda_2 du}\left(x_1 + \int_a^t e^{-\int_a^v \lambda_2 du}\sigma(v)dv\right)\right\| \\ &\qquad \leq e^{\Re\int_a^t \lambda_2 du}\int_t^b \varphi(v)e^{-\Re\int_a^v \lambda_2 du}dv,\end{aligned}$$

that is,

$$\begin{aligned}&\left\| g(t) - e^{t\lambda_2 - a\lambda_2}\left(x_1 + \int_a^t e^{-\int_a^v \lambda_2 du}\sigma(v)dv\right)\right\| \\ &\qquad \leq e^{\Re(t\lambda_2 - a\lambda_2)}\int_t^b \varphi(v)e^{-\Re\int_a^v \lambda_2 du}dv, \qquad (6.24)\end{aligned}$$

where $x_1 = \lim_{t\to b}\left(g(t)e^{-t\lambda_2 + a\lambda_2} - \int_a^t e^{\int_a^v \lambda_2 du}\sigma(v)dv\right)$. If

$$u(t) = x_1 e^{\int_a^t \lambda_2 du} + e^{\int_a^t \lambda_2 du}\int_a^t e^{-\int_a^v \lambda_2 du}\sigma(v)dv,$$

then

$$u'(t) = \lambda_2 e^{\int_a^t \lambda_2 du}\int_a^t e^{-\int_a^v \lambda_2 du}\sigma(v)dv + x_1\lambda_2 e^{\int_a^t \lambda_2 du} + \sigma(t)$$

and

$$\begin{aligned}u'(t) - \lambda_2 u(t) - \sigma(t) &= x_1\lambda_2 e^{\int_a^t \lambda_2 du} + \lambda_2 e^{\int_a^t \lambda_2 du}\int_a^t e^{-\int_a^v \lambda_2 du}\sigma(v)dv \\ &+\sigma(t) - x_1\lambda_2 e^{\int_a^t \lambda_2 du} - \lambda_2 e^{\int_a^t \lambda_2 du}\int_a^t e^{-\int_a^v \lambda_2 du}\sigma(v)dv - \sigma(t) = 0. \qquad (6.25)\end{aligned}$$

Let us consider $\psi(t) = e^{\Re(t\lambda_2 - a\lambda_2)} \int_t^b \varphi(v) e^{-\Re \int_a^v \lambda_2 du} dv$. *Since* $g(t) = y'(t) - \lambda_1 y(t)$, *then (6.24) becomes*

$$\|y'(t) - \lambda_1 y(t) - u(t)\| \leq \psi(t).$$

Again, by using Lemma 13, there exists a unique $x_2 \in X$ *such that*

$$\left\| y(t) - e^{\int_a^t \lambda_1 du} \left(x_2 + \int_a^t e^{-\int_a^v \lambda_1 du} .h(v) dv \right\| \leq e^{\Re \int_a^t \lambda_1 du} \int_a^t \psi(v) e^{-\Re \int_a^v \lambda_1 du} dv,$$

where $x_2 = \lim_{t \to b} \left(e^{-\int_a^t \lambda_1 du} y(t) - \int_a^t e^{-\int_a^v \lambda_1 du} .h(v) dv \right)$. *If* $y_0(t) = x_2 e^{\int_a^t \lambda_1 du}$
$+ e^{\int_a^t \lambda_1 du} \int_a^t e^{-\int_a^v \lambda_1 du} .h(v) dv$, *then* $y_0'(t) = \lambda_1 e^{\int_a^t \lambda_1 du} \int_a^t e^{-\int_a^v \lambda_1 du} h(v) dv + h(t)$
$+ x_2 \lambda_1 e^{\int_a^t \lambda_1 du}$ *and*

$$y_0'(t) - \lambda_1 y_0(t) - u(t) = 0.$$

So $u(t) = y_0'(t) - \lambda_1 y_0(t)$ *and* $u''(t) = y_0''(t) - \lambda_1 y_0'(t)$. *Using this fact in (6.25), we get*

$$y_0''(t) - \lambda_1 y_0'(t) - \lambda_2 \{ y_0'(t) - \lambda_1 y_0(t) \} - \sigma(t) = 0,$$

implies that

$$y_0''(t) + (-\lambda_1 - \lambda_2) y_0'(t) + \lambda_1 \lambda_2 y_0(t) - \sigma(t) = 0,$$

that is,

$$y_0''(t) + \alpha y_0'(t) + \beta y_0(t) = \sigma(t).$$

This completes the proof.

Corollary 28 *Let Let* $\varphi : I \to [0, \infty)$ *be an integrable function on* I. *Assume that* α,β *are complex numbers,* $f : I \to X$ *is continuous and integrable on* I. *If a twice continuously function* $y(t)$ *satisfies the inequality*

$$\|y''(t) + \alpha y'(t) + \beta y(t)\| \leq \varphi(t),$$

where $y : I \to X$ *is a twice continuously differentiable function,* X *is a real Banach space. Then (6.2) has the Hyers-Ulam-Rassias stability. Moreover, the approximating function is a real function.*

Proof 87 *Let* λ_1 *and* λ_2 *be the roots of the characteristic equation* $\lambda^2 + \alpha\lambda + \beta = 0$. *By using Theorem 50, (6.2) has the Hyers-Ulam stability. When the roots are real, then the approximating function is real. Let* λ_1 *and* λ_2 *be the roots of the characteristics equation which are complex numbers. Let* $\lambda_1 = p_1 + ip_2$ *and* $\lambda_2 = p_1 - ip_2$ *and* $\lim_{t \to b} y(t) = d_1$, $\lim_{t \to b} y'(t) = d_2$, *so* $\lim_{t \to b} g(t) =$

$d_2 - r_1 * d_1$, where p_1, p_2, d_1, d_2 are real number. After calculation, the approximating function is

$$\frac{1}{p_2}[d_1p_2\cos\{p_2(b-t)\} + (-d_2 + d_1p_1)\sin\{p_2(b-t)\}][\cosh\{p_1(b-t)\} - \sinh\{p_1(b-t)\}]$$

which is a real function. This completes the proof of the corollary.

Corollary 29 *Let $\varphi : I \to [0,\infty)$ be an integrable function on I. Assume that α,β are complex numbers, and $f : I \to X$ is continuous and integrable on I. If a twice continuously function $y(t)$ satisfies the inequality*

$$\|y''(t) + \alpha y'(t) + \beta y(t) - f(t)\| \leq \varphi(t),$$

where $y : I \to X$ is a twice continuously differentiable function and $f : I \to X$ is continuous and integrable on I, X is a real Banach space. Then (27) has the Hyers-Ulam-Rassias stability. Furthermore, the approximating function is a real function.

□

6.3 Notes

In Section 6.2, the contribution work [39] of Miura et al. have been kept for Hyers-Ulam stability of first order linear differential equations with holomorphic mappings. Section 6.3 provides the exclusive work of Y. Li and J. Huang [32].

Chapter 7

Hyers-Ulam Stability of Difference Equations

The objective of this chapter is to discuss the Hyers-Ulam stability of following difference equations

$$y_{n+2} + \alpha y_{n+1} + \beta y_n = 0 \tag{7.1}$$

and

$$y_{n+2} + \alpha y_{n+1} + \beta y_n = r_n \tag{7.2}$$

where $\alpha, \beta \in \mathbb{R}$ and r_n is a sequence of reals. Also, the Hyers-Ulam stability of

$$y_{n+2} - \alpha_n y_{n+1} + \beta_n y_n = r_n \tag{7.3}$$

is explained. Furthermore, the Hyers-Ulam stability of first order linear difference operator T_p defined by

$$(T_p u)(n) = \Delta u(n) - p(n)u(n),$$

is discussed on the Banach space $X = l_\infty$, where $p(n)$ is a sequence of reals.

7.1 Hyers-Ulam Stability of Second Order Difference Equations-I

This section deals with the Hyers-Ulam stability of (7.1) and (7.2).

Theorem 52 *Assume that the characteristic equation $m^2 + \alpha m + \beta = 0$ have two different positive roots. Then (7.1) has the Hyers-Ulam stability.*

Proof 88 *Let $\epsilon > 0$ and $y_n, n \in (a, b+1)$ be a solution of (7.1) satisfying the property*

$$|y_{n+2} + \alpha y_{n+1} + \beta y_n| \leq \epsilon.$$

Let λ and μ be the positive roots of the characteristic equation. For $n \in (a, b+1)$, define $g_n = y_{n+1} - \lambda y_n$. Then

$$g_{n+1} = y_{n+2} - \lambda y_{n+1}$$

and hence

$$\begin{aligned} |g_{n+1} - \mu g_n| &= |y_{n+2} - \lambda y_{n+1} - \mu y_{n+1} + \lambda\mu y_n| \\ &= |y_{n+2} - (\lambda + \mu) y_{n+1} + \lambda\mu y_n| \\ &= |y_{n+2} + \alpha y_{n+1} + \beta y_n| \leq \epsilon. \end{aligned}$$

Equivalently, g_n satisfies the relation

$$-\epsilon \leq g_{n+1} - \mu g_n \leq \epsilon. \tag{7.4}$$

Upon the choice of λ and μ, we have four possibilities viz.

i) $\lambda > 1$, $\mu > 1$; *ii)* $\lambda \leq 1$, $\mu \leq 1$; *iii)* $\lambda > 1$, $\mu \leq 1$; *iv)* $\lambda \leq 1$, $\mu > 1$.

Consider Case (i). Then (7.4) can be viewed as

$$-\epsilon\mu^{-(n+1)} \leq \mu^{-(n+1)}[g_{n+1} - \mu g_n] \leq \epsilon\mu^{-(n+1)},$$

that is,

$$-\epsilon\mu^{-(n+1)} \leq \Delta(\mu^{-n} g_n) \leq \epsilon\mu^{-(n+1)}. \tag{7.5}$$

Therefore for $n \in (a, b+1)$, it follows that

$$-\epsilon \sum_{j=n}^{b} \mu^{-(j+1)} \leq \sum_{j=n}^{b} \Delta(\mu^{-j} g_j) \leq \epsilon \sum_{j=n}^{b} \mu^{-(j+1)}.$$

which then implies that

$$\frac{-\epsilon\mu^{-n}}{\mu - 1} \leq \mu^{-(b+1)} g_{b+1} - \mu^{-n} g_n \leq \frac{\epsilon\mu^{-n}}{\mu - 1}.$$

Consequently,

$$-\epsilon_1 \leq \mu^{-(b-n+1)} g_{b+1} - g_n \leq \epsilon_1,$$

where $\epsilon_1 = \frac{\epsilon}{\mu-1}$. Let $z_n = \mu^{-(b-n+1)} g_{b+1}$. Then $z_{n+1} - \mu z_n = 0$. Now, $|g_n - z_n| \leq \epsilon_1$ implies that

$$-\epsilon_1 \leq y_{n+1} - \lambda y_n - z_n \leq \epsilon_1$$

and hence

$$-\epsilon_1 \lambda^{-(n+1)} \leq \lambda^{-(n+1)}[y_{n+1} - \lambda y_n - z_n] \leq \epsilon_1 \lambda^{-(n+1)}.$$

Proceeding as above, we obtain

$$-\epsilon_1 \frac{\lambda^{-n}}{\lambda - 1} \leq \lambda^{-(b+1)} y_{b+1} - \lambda^{-n} y_n - \sum_{j=n}^{b} \lambda^{-(j+1)} z_j \leq \epsilon_1 \frac{\lambda^{-n}}{\lambda - 1},$$

that is,

$$\frac{-\epsilon_1}{\lambda - 1} \leq \lambda^{-(b-n+1)} y_{b+1} - y_n - \lambda^n \sum_{j=n}^{b} \lambda^{-(j+1)} z_j \leq \frac{\epsilon_1}{\lambda - 1}.$$

If we denote

$$u_n = \lambda^{-(b-n+1)} y_{b+1} - \sum_{j=n}^{b} \lambda^{-(j-n+1)} z_j,$$

then $|u_n - y_n| \leq \frac{\epsilon_1}{\lambda-1} = \frac{\epsilon}{(\lambda-1)(\mu-1)}$. *It is easy to verify that* $u_{n+1} = \lambda u_n + z_n$ *and hence*

$$u_{n+2} - \lambda u_{n+1} = z_{n+1} = \mu z_n = \mu[u_{n+1} - \lambda u_n]$$

implies that

$$u_{n+2} + \alpha u_{n+1} + \beta u_n = 0.$$

Consequently, (7.1) *has the Hyers-Ulam stability with the stability constant* $K = \frac{1}{(\lambda-1)(\mu-1)}$.

Next, we consider Case(ii). Assume that there exist positive integers $M, L > 0$ *such that* $\mu M > 1$ *and* $\lambda L > 1$. *Using the same type of argument as in Case(i), we get* (7.5) *and hence*

$$-\epsilon \sum_{j=n}^{b} (\mu M)^{-(j+1)} M^{j+1} \leq \sum_{j=n}^{b} \Delta(\mu^{-j} g_j) \leq \epsilon \sum_{j=n}^{b} (\mu M)^{-(j+1)} M^{j+1}. \quad (7.6)$$

Let $M^* = max\{M^{n+1}, M^{n+2}, \cdots M^{b+1}\}$, $a \leq n < b+1$. *Then* (7.6) *becomes*

$$-\epsilon M^* \sum_{j=n}^{b} (\mu M)^{-(j+1)} \leq \mu^{-(b+1)} g_{b+1} - \mu^{-n} g_n \leq \epsilon M^* \sum_{j=n}^{b} (\mu M)^{-(j+1)},$$

that is,

$$M^* \frac{-\epsilon(\mu M)^{-n}}{(\mu M - 1)} \leq \mu^{-(b+1)} g_{b+1} - \mu^{-n} g_n \leq M^* \frac{\epsilon(\mu M)^{-n}}{(\mu M - 1)}.$$

Consequently,

$$\frac{-\epsilon M^*}{(\mu M - 1) M^a} \leq \mu^{-(b-n+1)} g_{b+1} - g_n \leq \frac{\epsilon M^*}{(\mu M - 1) M^a}.$$

The rest of the proof follows from Case (i). We note that the Hyers-Ulam stability constant is given by

$$K = \frac{\epsilon M^* L^*}{(\mu M - 1)(\lambda L - 1)(ML)^a},$$

where $L^* = max\{L^{n+1}, L^{n+2}, \ldots L^{b+1}\}$, $a \le n < b+1$.

Cases(iii) and (iv) follow from Cases(i) and (ii). Hence the proof of the theorem is complete.

Theorem 53 *Assume that the characteristic equation* $m^2 + \alpha m + \beta = 0$ *have two different positive roots. Furthermore, assume that*

$$|y_{n+2} + \alpha y_{n+1} + \beta y_n - r_n| \le \epsilon$$

holds. Then (7.2) has the Hyers-Ulam stability.

Proof 89 *Proceeding as in the proof of Theorem 7.2.1, we obtain*

$$\begin{aligned} |g_{n+1} - \mu g_n - r_n| &= |y_{n+2} - \lambda y_{n+1} - \mu y_{n+1} + \lambda\mu y_n - r_n| \\ &= |y_{n+2} - (\lambda + \mu) y_{n+1} + \lambda\mu y_n - r_n| \\ &= |y_{n+2} + \alpha y_{n+1} + \beta y_n - r_n| \le \epsilon. \end{aligned}$$

Equivalently, g_n *satisfies the relation*

$$-\epsilon \le g_{n+1} - \mu g_n - r_n \le \epsilon.$$

Similar to Theorem 7.2.1, we have four possibilities upon the choices of λ *and* μ. *We consider Case (i) only. Other cases can similarly be dealt with the cases in Theorem 7.2.1. Hence, similar to (7.5), we have*

$$-\epsilon\mu^{-(n+1)} \le \Delta(\mu^{-n} g_n) - \mu^{-(n+1)} r_n \le \epsilon\mu^{-(n+1)},$$

and we let $z_n = \mu^{-(b-n+1)} g_{b+1} - \mu^n \sum_{j=n}^{b} \mu^{-(j+1)} r_j$. *Therefore,* z_n *satisfies*

$$z_{n+1} - \mu z_n - r_n = 0,$$

and $|g_n - z_n| \le \epsilon_1$. *Using the same type of argument as in Theorem 7.2.1, we can show that there exists*

$$u_n = \lambda^{-(b-n+1)} y_{b+1} - \sum_{j=n}^{b} \lambda^{-(j-n+1)} z_j$$

such that $|u_n - y_n| \le \frac{\epsilon}{(\lambda-1)(\mu-1)}$ *and* u_n *satisfies*

$$u_{n+2} + \alpha u_{n+1} + \beta u_n - r_n = 0.$$

This completes the proof of the theorem.

7.2 Hyers-Ulam Stability of Second Order Difference Equations-II

In this section, the Hyers-Ulam stability of (7.3) is presented. For this we need the following result:

Lemma 14 *Assume that $0 \leq p_n \leq p < \infty$ for every n. Then for $n \in (a, b+1)$, the equation*

$$y_{n+1} - p_n y_n - r_n = 0 \tag{7.7}$$

has the Hyers-Ulam stability.

Proof 90 *Let $\epsilon > 0$ and $y_n, n \in (a, b+1)$ be a solution of (7.7) satisfying the property*

$$|y_{n+1} - p_n y_n - r_n| \leq \epsilon,$$

that is,

$$-\epsilon \leq y_{n+1} - p_n y_n - r_n \leq \epsilon.$$

Therefore,

$$-\epsilon\left(\prod_{i=a}^{n} p_i\right)^{-1} \leq [y_{n+1} - p_n y_n - r_n]\left(\prod_{i=a}^{n} p_i\right)^{-1} \leq \epsilon\left(\prod_{i=a}^{n} p_i\right)^{-1}$$

implies that

$$-\epsilon\left(\prod_{i=a}^{n} p_i\right)^{-1} \leq \Delta[y_n\left(\prod_{i=a}^{n-1} p_i\right)^{-1}] - r_n\left(\prod_{i=a}^{n} p_i\right)^{-1} \leq \epsilon\left(\prod_{i=a}^{n} p_i\right)^{-1}.$$

Summing the above inequality from a to $n-1$, we obtain

$$-\epsilon\sum_{j=a}^{n-1}\left(\prod_{i=a}^{j} p_i\right)^{-1} \leq \sum_{j=a}^{n-1}\Delta[y_j\left(\prod_{i=a}^{j-1} p_i\right)^{-1}] - \sum_{j=a}^{n-1} r_j\left(\prod_{i=a}^{j} p_i\right)^{-1} \leq \epsilon\sum_{j=a}^{n-1}\left(\prod_{i=a}^{j} p_i\right)^{-1}$$

and hence

$$-\epsilon\sum_{j=a}^{n-1}\left(\prod_{i=a}^{j} p_i\right)^{-1} \leq y_n\left(\prod_{i=a}^{n-1} p_i\right)^{-1} - y_a - \sum_{j=a}^{n-1} r_j\left(\prod_{i=a}^{j} p_i\right)^{-1} \leq \epsilon\sum_{j=a}^{n-1}\left(\prod_{i=a}^{j} p_i\right)^{-1}, \tag{7.8}$$

where we have used the sign convention $\prod_{i=a}^{a-1} p_i = 1$. We consider two cases, *viz.* $p < 1$ *and* $p \geq 1$. *Consider the former case. Using the fact*

$$\sum_{j=a}^{n-1}\left(\prod_{i=a}^{j} p_i\right)^{-1} = \frac{1}{p_a} + \frac{1}{p_a p_{a+1}} + \cdots + \frac{1}{p_a p_{a+1} \cdots p_{n-1}} \leq \frac{1}{1-p}\left(\prod_{i=a}^{n-1} p_i\right)^{-1}$$

in (7.8), *we obtain that*

$$-\frac{\epsilon}{1-p}\left(\prod_{i=a}^{n-1} p_i\right)^{-1} \leq y_n \left(\prod_{i=a}^{n-1} p_i\right)^{-1} - y_a - \sum_{j=a}^{n-1} r_j \left(\prod_{i=a}^{j} p_i\right)^{-1} \leq \frac{\epsilon}{1-p}\left(\prod_{i=a}^{n-1} p_i\right)^{-1},$$

that is,

$$-\frac{\epsilon}{1-p} \leq y_n - y_a\left(\prod_{i=a}^{n-1} p_i\right) - \left(\prod_{i=a}^{n-1} p_i\right)\sum_{j=a}^{n-1} r_j \left(\prod_{i=a}^{j} p_i\right)^{-1} \leq \frac{\epsilon}{1-p}.$$

For $n \in (a, b+1)$, *if we define*

$$z_n = y_a\left(\prod_{i=a}^{n-1} p_i\right) + \left(\prod_{i=a}^{n-1} p_i\right)\sum_{j=a}^{n-1} r_j \left(\prod_{i=a}^{j} p_i\right)^{-1},$$

then it is easy to verify that $z_{n+1} - p_n z_n - r_0 = 0$ *and hence* $|y_n - z_n| \leq \frac{\epsilon}{1-p}$ *implies that* (7.7) *has the Hyers-Ulam stability with stability constant* $K = \frac{1}{1-p}$. *For the latter case, we have*

$$\sum_{j=a}^{n-1}\left(\prod_{i=a}^{j} p_i\right)^{-1} = \frac{1}{p_a} + \frac{1}{p_a p_{a+1}} + \cdots + \frac{1}{p_a p_{a+1} \cdots p_{n-1}} \leq \frac{p^n - 1}{p-1}\left(\prod_{i=a}^{n-1} p_i\right)^{-1} \leq \frac{p^b - 1}{p-1}\left(\prod_{i=a}^{n-1} p_i\right)^{-1}.$$

Using the same type of argument as in the former case, we obtain that $|y_n - z_n| \leq \frac{p^b-1}{p-1}\epsilon$. *This completes the proof of the lemma.*

Theorem 54 *Assume that* $\alpha_n, \beta_n > 0$ *for every* n. *Furthermore, assume that* c_n *is a particular solution of*

$$u_{n+1}u_n - \alpha_n u_n + \beta_n = 0 \tag{7.9}$$

such that $0 < c_n \leq c < \infty$ *and* $d_n = \frac{\beta_n}{c_n} \leq d < \infty$, *for* $n \in (a, b+1)$. *Then* (7.3) *has the Hyers-Ulam stability.*

Proof 91 *Let $\epsilon > 0$ and $y_n, n \in (a, b+1)$ be a solution of (7.3) satisfying the property*

$$|y_{n+2} - \alpha_n y_{n+1} + \beta_n y_n - r_n| \leq \epsilon.$$

For $n \in (a, b+1)$, define $v_n = y_{n+1} - c_n y_n$. Then

$$\begin{aligned}|v_{n+1} - d_n v_n - r_n| &= |y_{n+2} - (d_n + c_{n+1})y_{n+1} + c_n d_n y_n - r_n| \\ &= |y_{n+2} - \alpha_n y_{n+1} + \beta_n y_n - r_n| \leq \epsilon,\end{aligned}$$

where we have used the fact that c_n is a particular solution of (7.9). Upon the choice of c and d, we undertake four possibilities viz.

i) $c < 1$, $d < 1$; *ii)* $c < 1$, $d \geq 1$; *iii)* $c \geq 1$, $d < 1$; *iv)* $c \geq 1$, $d \geq 1$.

We consider Case (i) and the other cases can similarly be dealt with. It follows from Lemma 7.3.1 that

$$v_{n+1} - d_n v_n - r_n = 0$$

has the Hyers-Ulam stability with the property $|v_n - w_n| \leq \frac{\epsilon}{1-d}$, where

$$w_n = v_a \left(\prod_{i=a}^{n-1} d_i\right) + \left(\prod_{i=a}^{n-1} d_i\right) \sum_{j=a}^{n-1} r_j \left(\prod_{i=a}^{j} d_i\right)^{-1}.$$

Using v_n in $|v_n - w_n| \leq \frac{\epsilon}{1-d}$, we obtain that

$$|y_{n+1} - c_n y_n - w_n| \leq \frac{\epsilon}{1-d}$$

for $n \in (a, b+1)$. Again by Lemma 7.3.1,

$$y_{n+1} - c_n y_n - w_n = 0$$

has the Hyers-Ulam stability with the property $|y_n - z_n| \leq \frac{\epsilon}{(1-d)(1-c)}$, where

$$z_n = y_a \left(\prod_{i=a}^{n-1} c_i\right) + \left(\prod_{i=a}^{n-1} c_i\right) \sum_{j=a}^{n-1} w_j \left(\prod_{i=a}^{j} c_i\right)^{-1}.$$

Consequently, (7.3) has the Hyers-Ulam stability with the stability constant $K = \frac{1}{(1-c)(1-d)}$. Hence the proof of the theorem is complete.

Example 15 *Consider the difference equation*

$$y_{n+2} - \frac{1}{n+2} y_{n+1} + \frac{1}{(n+1)(n+2)} y_n = \frac{n^2 - n + 1}{n(n+1)(n+2)} \tag{7.10}$$

on the interval $I = (1, \infty)$. *Let* $c_n = \frac{1}{n+1}, n \in I$ *be a particular solution of*

$$u_{n+1}u_n - \frac{2}{n+2}u_n + \frac{1}{(n+1)(n+2)} = 0.$$

By Theorem 7.3.2, $d_n = \frac{1}{n+2} \leq d < 1$, $\prod_{i=1}^{n-1} d_i = \frac{2}{(n+1)!}$ *and* $\prod_{i=1}^{n-1} c_i = \frac{1}{(n)!}$. *It is easy to verify that* $w_n \geq \frac{1}{(n+1)!}$ *and* $z_n \geq \frac{1}{(n)!}$ *for* $n \in I$. *Indeed,* $y_n = \frac{1}{n}$ *is a solution of* (7.10) *and* $|y_n - z_n| \leq \frac{\epsilon}{(1-c)(1-d)}$. *Hence* (7.10) *has the Hyers-Ulam stability.*

7.3 Hyers-Ulam Stability First Order Difference Operators

Let $X = l_\infty$ be the Banach space of all real valued functions $u(n)$ defined for $n \geq 0$. Let $D(I, X)$ be the linear space of all $X-$valued functions on an open interval $I = (a, b+1) \subset \mathbb{N}(0) = \{0, 1, 2, ...\}$, $a < b$. We define

$$\|f\|_\infty = \sup\{\|f(n)\|, n \in I\}$$

for every $f \in D(I, X)$. Define the linear difference operator $T_p : D(I, X) \to D(I, X)$ by

$$(T_p u)(n) = \Delta u(n) - p(n)u(n), \ \forall u \in D(I, X), \forall\, n \in I. \tag{7.11}$$

We notice that T_p is onto. Indeed, for every $v \in D(I, X)$

$$u(n) = \prod_{i=n_0}^{n-1}(1 + p(i)) \sum_{s=n_0}^{n-1} v(s) \left(\prod_{i=n_0}^{s}(1 + p(i))\right)^{-1}$$

satisfies $T_p = v$. Conversely, the general solution of $T_p = v$ is of the form

$$u(n) = \prod_{i=n_0}^{n-1}(1 + p(i)) \left[x_0 + \sum_{s=n_0}^{n-1} v(s) \left(\prod_{i=n_0}^{s}(1 + p(i))\right)^{-1}\right]$$

for every $n_0, n \in I$ and $x_0 \in X$ is an arbitrary element.

Definition 11 *We say that the difference operator* T_p *has the Hyers-Ulam stability, if there exists a constant* $K \geq 0$ *with the property:* For every $\epsilon \geq 0$ and $u, v \in D(I, X)$ satisfying $\|T_p u - v\|_\infty \leq \epsilon$ there exists $u_0 \in D(I, X)$ such that $T_p u_0 = v$ and $\|u - u_o\|_\infty \leq K\epsilon$. *We call such* K *a HUS constant for* T_p. *If, in addition, minimum of all such* K*'s exists, then we call it the HUS constant for* T_p.

We use the following lemma due to [57] in our next discussion:

Lemma 15 *Let C be a symmetric set, that is, $C = -C$ in a Banach space B. For each $y \in B$, we have*

$$\sup_{x\in C} \|y + x\| \geq \sup_{x\in C} \|x\|.$$

In this section, we discuss the necessary and sufficient conditions for Hyers-Ulam stability of the operator T_p followed by (7.11) on the Banach space $X = l_\infty$. We use the following notions for our use in the sequel:

$$1 + p(n) \neq 0, \quad P(n) = \left(\prod_{i=0}^{n-1}(1 + p(i))\right)^{-1}, \; n \in \mathbb{N}(0),$$

$$\alpha_p = \sup_{n\geq 0} \frac{1}{|P(n)|} \sum_{m=n}^{\infty} |P(m+1)|, \quad \beta_p = \sup_{n\geq 0} \frac{1}{|P(n)|} \sum_{m=0}^{n-1} |P(m+1)|.$$

We use the sign convention $\left(\prod_{i=0}^{n-1}(1 + p(i))\right) = 1$ for $n - 1 < 0$.

Theorem 55 *Let $T_p : D(\mathbb{N}(0), X) \to D(\mathbb{N}(0), X)$ be the linear operator defined by*

$$(T_p u)(n) = \triangle u(n) - p(n)u(n), \; \forall u \in D(\mathbb{N}(0), X), \forall\, n \in \mathbb{N}(0). \tag{7.12}$$

If $\inf_{n\geq 0} |P(n)| = 0$, then T_p has the Hyers-Ulam stability with HUS constant α_p if and only if $\alpha_p < \infty$.

Proof 92 *Let $\epsilon \geq 0$ and $u, v \in D(\mathbb{N}(0), X)$ satisfy $\|T_p u - v\|_\infty \leq \epsilon$. Set $w = T_p u - v$. Then $\|w\|_\infty \leq \epsilon$ and $T_p u = v + w$ implies that*

$$\begin{aligned}
u(n) &= \prod_{i=0}^{n-1}(1 + p(i)) \left[u(0) + \sum_{s=0}^{n-1}(v + w)\left(\prod_{i=0}^{s}(1 + p(i))\right)^{-1}\right] \\
&= \prod_{i=0}^{n-1}(1 + p(i)) \sum_{s=0}^{n-1} v(s) \left(\prod_{i=0}^{s}(1 + p(i))\right)^{-1} \\
&+ \prod_{i=0}^{n-1}(1 + p(i)) \left[u(0) + \sum_{s=0}^{n-1} w(s)\left(\prod_{i=0}^{s}(1 + p(i))\right)^{-1}\right] \\
&= \frac{1}{P(n)} \sum_{s=0}^{n-1} v(s)P(s+1) + \frac{1}{P(n)}\left[u(0) + \sum_{s=0}^{n-1} w(s)P(s+1)\right],
\end{aligned} \tag{7.13}$$

and $\sum_{s=0}^{n-1} w(s)P(s+1) \in X$ exists for every $n \in \mathbb{N}(0)$.

Now, we consider the case when $\alpha_p < \infty$. For each $n \in \mathbb{N}(0)$, it is easy to see that

$$\begin{aligned} u(n) =& \frac{1}{P(n)} \sum_{s=0}^{n-1} v(s)P(s+1) + \frac{1}{P(n)} \left[u(0) + \sum_{s=0}^{\infty} w(s)P(s+1) \right] \\ &- \frac{1}{P(n)} \sum_{s=n}^{\infty} w(s)P(s+1). \end{aligned}$$

If we put $x_0 = u(0) + \sum_{s=0}^{\infty} w(s)P(s+1)$ and

$$u_0(n) = \frac{1}{P(n)} \left[x_0 + \sum_{s=0}^{n-1} v(s)P(s+1) \right],$$

then $T_p u_0 = v$ and

$$u(n) = u_0(n) - \frac{1}{P(n)} \sum_{s=n}^{\infty} w(s)P(s+1).$$

Therefore,

$$\|u(n) - u_0(n)\| = \frac{1}{|P(n)|} \| \sum_{s=n}^{\infty} w(s)P(s+1)\|$$

implies that

$$\|u - u_0\|_\infty \le \sup_{n \ge 0} \frac{\epsilon}{|P(n)|} \sum_{s=n}^{\infty} |P(s+1)| \le \epsilon \alpha_p.$$

Hence, T_p has the Hyers-Ulam stability with HUS constant α_p. We claim that u_0 is determined uniquely. If not, let $u_1, u_2 \in D(\mathbb{N}(0), X)$ be such that

$$T_p u_i = v \text{ and } \|u - u_i\|_\infty \le M_i < \infty, \quad (i = 1, 2).$$

Hence for $T_p u_i = v$, we can find x_i for $i = 1, 2$ such that

$$u_i(n) = \frac{1}{P(n)} \left[x_i + \sum_{s=0}^{n-1} v(s)P(s+1) \right], \ \forall\, n \in \mathbb{N}(0).$$

Therefore, it follows that

$$\begin{aligned} \|x_1 - x_2\| = |P(n)| \|u_1(n) - u_2(n)\| &\le |P(n)| \|u_1 - u_2\|_\infty \\ &\le |P(n)|(M_1 + M_2), \ \forall\, n \in \mathbb{N}(0), \end{aligned}$$

that is, $\|x_1 - x_2\| \to 0$ due to $\inf_{n \ge 0} |P(n)| = 0$. Consequently, $u_1 = u_2$.

Conversely, let's fix $x_0 \in X$ such that $\|x_0\| = 1$. Set $v(n) = \frac{|P(n+1)|}{P(n+1)} x_0$ for *every* $n \in \mathbb{N}(0)$. *Then for* $v \in D(\mathbb{N}(0), X)$ *we can find* $u \in D(\mathbb{N}(0), X)$ *such that*

$$u(n) = \frac{1}{P(n)} \sum_{s=0}^{n-1} v(s)P(s+1) = \frac{1}{P(n)} \sum_{s=0}^{n-1} |P(s+1)|x_0$$

for which $T_p u = v$ *and hence* $\|T_p u\|_\infty = \|v\|_\infty = 1$. *Let* K *be an arbitrary HUS constant for* T_p. *So, we can find* $u^* \in D(\mathbb{N}(0), X)$ *such that*

$$T_p u^* = 0 \text{ and } \|u - u^*\|_\infty \leq K.$$

It is easy to verify that $u^* = \frac{x_1}{P(n)}$ *for every* $n \in \mathbb{N}(0)$, *where* $x_1 = u^*(0) \in X$. *Therefore,*

$$\begin{aligned} \|\sum_{s=0}^{n-1} |P(s+1)|x_0 - x_1\| &= |P(n)| \|\frac{1}{P(n)} \sum_{s=0}^{n-1} |P(s+1)|x_0 - u^*\| \\ &\leq K|P(n)|, \ n \in \mathbb{N}(0). \end{aligned} \tag{7.14}$$

As $\inf_{n\geq 0} |P(n)| = 0$, *we can find a strictly monotonic increasing set of values* $\{n_j\}_{j\in\mathbb{N}} \subset \mathbb{N}(0)$ *such that*

$$n_j \to \infty \text{ as } j \to \infty \text{ and } |P(n_j)| < \frac{1}{j}, \ j \in \mathbb{N}.$$

From (7.14), *it follows that*

$$\left| \| \sum_{s=0}^{n_j-1} |P(s+1)|x_0\| - \|x_1\| \right| \leq \frac{K}{j},$$

that is,

$$\sum_{s=0}^{n_j-1} |P(s+1)| < \infty \text{ as } j \to \infty.$$

Also, from (7.13) *it is immediate that* $\sum_{s=0}^{\infty} |P(s+1)|x_0 = x_1$. *Consequently,*

$$\begin{aligned} \sum_{s=n}^{\infty} |P(s+1)| &= \|\sum_{s=0}^{n-1} |P(s+1)|x_0 - x_1\| \\ &= \|\sum_{s=0}^{n-1} |P(s+1)|x_0 - \sum_{s=0}^{\infty} |P(s+1)|x_0\| \leq K|P(n)| \end{aligned}$$

implies that $\alpha_p \leq K < \infty$. *Since* K *is an arbitrary HUS constant, then* α_p *itself is the HUS constant for* T_p. *This completes the proof of the theorem.*

Remark 8 *We predict that β_p could be infinity when α_p is finite. In this case, $\sum |P(m)| < \infty$. Ultimately, $\inf_{n\geq 0} |P(n)| = 0$. So, we can find a strictly monotonic increasing set of values $\{n_j\}_{j\in\mathbb{N}} \subset \mathbb{N}(0)$ such that*

$$n_j \to \infty \text{ as } j \to \infty \text{ and } |P(n_j)| < \frac{1}{j},\ j \in \mathbb{N}.$$

Consequently,

$$\beta_p \geq \frac{1}{|P(n_j)|} \sum_{m=0}^{n_j-1} |P(m+1)| > j \sum_{m=0}^{n_j-1} |P(m+1)| \to \infty \text{ as } j \to \infty.$$

Theorem 56 *Let $T_p : D(\mathbb{N}(0), X) \to D(\mathbb{N}(0), X)$ be the linear operator defined by (7.12). If $\inf_{n\geq 0} |P(n)| > 0$, then T_p has the Hyers-Ulam stability with HUS constant β_p if and only if $\beta_p < \infty$.*

Proof 93 *We proceed as in the proof of Theorem 7.4.3 to obtain (7.13). If we denote*

$$u_3(n) = \frac{1}{P(n)} \left[u(0) + \sum_{s=0}^{n-1} v(s)P(s+1) \right],$$

then $T_p u_3 = v$ and

$$u(n) = u_3(n) + \frac{1}{P(n)} \sum_{s=0}^{n-1} w(s)P(s+1)$$

implies that

$$\|u - u_3\|_\infty \leq \sup_{n\geq 0} \frac{\epsilon}{|P(n)|} \sum_{s=0}^{n-1} |P(s+1)| \leq \epsilon\beta_p. \tag{7.15}$$

Hence, T_p has the Hyers-Ulam stability with HUS constant β_p.

Assume that $\inf_{n\geq 0} |P(n)| > 0$. Proceeding as in the converse part of Theorem 7.4.3, we have $u^*(n) = \frac{x_1}{P(n)}$ for $n \in \mathbb{N}(0)$. Hence,

$$\sup_{n\geq 0} \|u_0(n)\| \leq \frac{\|x_1\|}{\inf_{n\geq 0} |P(n)|} < \infty$$

implies that $\|x_1\| \leq \|u_0\|_\infty |P(n)|$ for every $n \in \mathbb{N}(0)$. Therefore,

$$\sum_{m=0}^{n-1} |P(m+1)| = \left\| \sum_{m=0}^{n-1} |P(m+1)| x_0 \right\| \leq (K + \|u_0\|_\infty)|P(n)|,$$

that is, $\beta_p \leq (K + \|u_0\|_\infty) < \infty$. Since K is an arbitrary HUS constant, then β_p is the HUS constant for T_p. Hence, the theorem is proved.

Remark 9 *In Theorem 7.4.3, we have seen that the uniqueness is true when $\alpha_p < \infty$. However, the same may not be true for the case when $\beta_p < \infty$. In other words, if K is an arbitrary constant with $\beta_p < K$, then for every $\epsilon > 0$ and $u, v \in D(\mathbb{N}(0), X)$ satisfying $\|T_p u - v\|_\infty \leq \epsilon$, we can find infinitely many $w \in D(\mathbb{N}(0), X)$ such that $T_p w = v$ and $\|u - w\|_\infty \leq K\epsilon$.*

Indeed, T_p has the Hyers-Ulam stability due to Remark 7.4.1 and (7.15) holds. Due to Theorem 7.4.4, let's put $\sigma = \inf_{n \geq 0} |P(n)|$. For each $x \in X$ with $\|x - u_3\|_\infty \leq \sigma\epsilon(K - \beta_p)$, we can define $u_x \in D(\mathbb{N}(0), X)$ by

$$u_x(n) = \frac{1}{P(n)} \left[x + \sum_{s=0}^{n-1} v(s) P(s+1) \right]$$

such that $T_p u_x = v$ and

$$\begin{aligned} \|u(n) - u_x(n)\| &\leq \|u(n) - u_3(n)\| + \|u_3(n) - u_x(n)\| \\ &\leq \epsilon\beta_p + \frac{1}{|P(n)|} \|x - u_3(n)\| \\ &\leq \epsilon\beta_p + \frac{\sigma\epsilon}{|P(n)|}(K - \beta_p) \leq K\epsilon \end{aligned}$$

for every $n \in \mathbb{N}(0)$. Hence, continuing in this way we can find many $w \in D(\mathbb{N}(0), X)$ such that $T_p w = v$ and $\|u - w\|_\infty \leq K\epsilon$.

Remark 10 *Since the uniqueness doesn't hold in case when $\beta_p < \infty$, then the simultaneous question is whether the infimum of all HUS constants, that is,*

$$\inf_{x \in X} \sup_{\substack{w \in D(\mathbb{N}(0), X) \\ \|w\|_\infty \leq 1}} \sup_{n \geq 0} \left\| \frac{1}{P(n)} \left[x + \sum_{s=0}^{n-1} w(s) P(s+1) \right] \right\|$$

if it exists for T_p is a HUS constant or not. Indeed, if we denote

$$L_{T_p} = \inf_{x \in X} \sup_{\substack{w \in D(\mathbb{N}(0), X) \\ \|w\|_\infty \leq 1}} \sup_{n \geq 0} \left\| \frac{1}{P(n)} \left[x + \sum_{s=0}^{n-1} w(s) P(s+1) \right] \right\|$$

and

$$L_0(x) = \sup_{\substack{w \in D(\mathbb{N}(0), X) \\ \|w\|_\infty \leq 1}} \sup_{n \geq 0} \left\| \frac{1}{P(n)} \left[x + \sum_{s=0}^{n-1} w(s) P(s+1) \right] \right\|,$$

then it is enough to verify that $L_{T_p} = \inf_{x \in X} L_0(x)$.

If $L_{T_p} = \infty$, then there is nothing to verify. Assume that $L_{T_p} < \infty$. Let K be an arbitrary HUS constant for T_p. Then for any $w \in D(\mathbb{N}(0), X)$ with

$\|w\|_\infty \leq 1$, *there exists* $u_0 \in D(\mathbb{N}(0), X)$ *such that* $T_p u_0 = w$ *and* $\|u_0\|_\infty \leq K$ *with*

$$u_0(n) = \frac{1}{P(n)}\left[x_0 + \sum_{s=0}^{n-1} w(s)P(s+1)\right]$$

for some $x_0 \in X$. *Indeed,* $L_0(x) \leq K$. *Since* K *is an arbitrary HUS constant for* T_p, *then it follows that* $L_{T_p} \geq \inf_{x\in X} L_0(x)$. *Conversely, we show that* $L_{T_p} \leq \inf_{x\in X} L_0(x)$. *We may assume that* $\inf_{x\in X} L_0(x) < \infty$.

Here, we assert that $L_0(x)$ *is a HUS constant for* T_p, *that is, for any* $\epsilon > 0$ *and* $u, v \in D(\mathbb{N}(0), X)$ *with* $\|T_p u - v\| \leq \epsilon$, *there exists* $u_0 \in D(\mathbb{N}(0), X)$ *such that* $T_p u_0 = v$ *and* $\|u - u_0\|_\infty \leq \epsilon L_0(x)$.

If we put $\epsilon w = T_p u - v$ *for* $u, v \in D(\mathbb{N}(0), X)$, *then* $\|w\|_\infty \leq 1$. *Hence for* $\epsilon w + v = T_p u$ *and for any arbitrary* $x_1 \in X$, *we have*

$$u(n) = \frac{1}{P(n)}\left[x_1 + \epsilon\sum_{s=0}^{n-1} w(s)P(s+1) + \sum_{s=0}^{n-1} v(s)P(s+1)\right]$$

for any $n \in \mathbb{N}(0)$. *Let*

$$u_0(n) = \frac{1}{P(n)}\left[x_1 - \epsilon x + \sum_{s=0}^{n-1} v(s)P(s+1)\right], \ n \in \mathbb{N}(0).$$

Then $u_0 \in D(\mathbb{N}(0), X)$ *and* $T_p u_0 = v$. *Consequently,*

$$\begin{aligned}\|u - u_0\|_\infty &= \sup_{n\geq 0}\|u(n) - u_0(n)\| \\ &= \sup_{n\geq 0}\left\|\frac{\epsilon}{P(n)}\left[x + \sum_{s=0}^{n-1} w(s)P(s+1)\right]\right\| \leq \epsilon L_0(x)\end{aligned}$$

due to the fact that $\|w\|_\infty \leq 1$. *Hence,* $L_0(x)$ *is a HUS constant for* T_p. *Thus,* $L_{T_p} \leq L_0(x)$. *Since* $x \in X$ *is arbitrary, then it follows that* $L_{T_p} \leq \inf_{x\in X} L_0(x)$.

Theorem 57 *Let* $T_p : D(\mathbb{N}(0), X) \to D(\mathbb{N}(0), X)$ *be the linear operator defined by* (7.12). *If* $\beta_p < \infty$, *then* T_p *has the Hyers-Ulam stability with HUS constant* L_{T_p}.

Proof 94 *Suppose that* $\beta_p < \infty$. *Then by Remark 7.4.1,* T_p *has the Hyers-Ulam stability with HUS constant* β_p. *Because the uniqueness doesn't hold in case when* $\beta_p < \infty$ *due to Remark 7.4.2 and* $L_{T_p} < \infty$ *exists due to Remark 7.4.3, then it is sufficient to show that* $L_{T_p} = \beta_p$. *By definition,* $L_{T_p} \leq \beta_p$. *Hence, we need to show that* $L_{T_p} \geq \beta_p$ *only. Define a linear operator* $S : D(\mathbb{N}(0), X) \to D(\mathbb{N}(0), X)$ *by*

$$(Sw)(n) = \frac{1}{P(n)}\sum_{s=0}^{n-1} w(s)P(s+1), \ \forall\, n \in \mathbb{N}(0), \ \ w \in D(\mathbb{N}(0), X).$$

Then for all $w \in D(\mathbb{N}(0), X)$,

$$\|Sw\|_\infty = \sup_{n\geq 0} \|(Sw)(n)\| = \sup_{n\geq 0} \left\| \frac{1}{P(n)} \sum_{s=0}^{n-1} w(s)P(s+1) \right\|$$
$$\leq \sup_{n\geq 0} \frac{\|w\|_\infty}{|P(n)|} \sum_{s=0}^{n-1} |P(s+1)| = \beta_p \|w\|_\infty < \infty.$$

Hence, S *is a bounded linear operator with* $\|S\| \leq \beta_p$*. Moreover, if* x_0 *is a unit element of* X *and* $u_0 = \frac{|P(n)|}{P(n)} x_0$ *for* $n \in \mathbb{N}(0)$*, then* $u_0 \in D(\mathbb{N}(0), X)$ *and* $\|u_0\|_\infty = 1$*. Consequently,* $\|Su_0\|_\infty = \beta_p$ *and hence* $\|S\| \geq \beta_p$*. Therefore,* $\|S\| = \beta_p$*.*

Since $|P(0)| = 1$*, then we can find* $n_1 > 0$ *such that* $|P(n)| \geq \frac{1}{2}$ *for* $n \geq n_1 + 1$*. Thus,*

$$\beta_p \geq \frac{1}{|P(n)|} \sum_{s=0}^{n-1} |P(s+1)| \geq \frac{1}{|P(n)|} \sum_{s=n_1}^{n-1} |P(s+1)| \geq \frac{(n-n_1)}{2|P(n)|},$$

that is, $\frac{1}{|P(n)|} \leq 2\beta_p < \infty$*. Therefore, if we choose* $x \in X$ *arbitrary, then it follows that* $\frac{x}{|P(n)|} \in D(\mathbb{N}(0), X)$*. We notice that the set* $(\{w \in D(\mathbb{N}(0), X) : \|w\|_\infty \leq 1\})$ *is a symmetric set of* $D(\mathbb{N}(0), X)$*. Applying Lemma 7.4.2, we obtain*

$$\sup_{\substack{w\in D(\mathbb{N}(0),X)\\ \|w\|_\infty\leq 1}} \sup_{n\geq 0} \left\| \frac{1}{P(n)} \left[x + \sum_{s=0}^{n-1} w(s)P(s+1) \right] \right\|$$
$$= \sup_{\substack{w\in D(\mathbb{N}(0),X)\\ \|w\|_\infty\leq 1}} \sup_{n\geq 0} \left\| \frac{x}{P(n)} + (Sw)(n) \right\|$$

$$= \sup_{\substack{w\in D(\mathbb{N}(0),X)\\ \|w\|_\infty\leq 1}} \left\| \frac{x}{P} + (Sw) \right\|_\infty \geq \sup_{\substack{w\in D(\mathbb{N}(0),X)\\ \|w\|_\infty\leq 1}} \|(Sw)\|_\infty = \|S\|.$$

which holds for all $x \in X$*. Ultimately,*

$$\inf_{x\in X} \sup_{\substack{w\in D(\mathbb{N}(0),X)\\ \|w\|_\infty\leq 1}} \sup_{n\geq 0} \left\| \frac{1}{P(n)} \left[x + \sum_{s=0}^{n-1} w(s)P(s+1) \right] \right\| \geq \|S\| = \beta_p.$$

This completes the proof of the theorem.

Example 16 *Consider*

$$(T_p u)(n) = \triangle u(n) - (1 + (-1)^n) u(n)$$

such that $1+p(n)=2+(-1)^n$ *and* $P(n)=\left(\prod_{i=0}^{n-1}(2+(-1)^i)\right)^{-1}$. *Indeed,*

$$\begin{aligned}
&\frac{1}{P(n)}\sum_{m=n}^{\infty}P(m+1)\\
&=\frac{1}{(2+(-1)^n)}\left[1+\frac{1}{(2+(-1)^n)}+\frac{1}{(2+(-1)^n)(2-(-1)^n)}+\cdots\right]\\
&\leq 2\left[1+\frac{1}{1.3}+\frac{1}{1^2.3^2}+\cdots\right]=3
\end{aligned}$$

implies that $\alpha_p \leq 3$. *Hence by Theorem7.4.3,* T_p *has the Hyers-Ulam stability with HUS constant* $\alpha_p \leq 3$.

□

7.4 Notes

In Sections 7.2 and 7.3, the contribution work [60] of A. K. Tripathy have been kept for Hyers-Ulam stability of second order linear difference equations. Section 7.4 provides the exclusive work of Tripathy et al. [61] for first order difference operators on Banach space.

Bibliography

[1] C. Alsina, R. Ger, On some inequalities and stability results related to the exponential function, J. Ineq. Appl., **2**(1998), 373–380.

[2] T. Aoki, On the stability of the linear transformation in Banach spaces, J. Math. Soc. Japan, **2**(1950), 1–2.

[3] T. M. Apostol; *Mathematical Analysis*, Addison Wesley Publishing Company, Indian Student Edition, 1997.

[4] J. H. Bae, Y. S. Jung, The Hyers-Ulam stability of the quadratic functional equations on abelian groups, Bull. Korean Math. Soc., **39**(2002), 199–209.

[5] F. Bojor, Note on the stability of first order linear differential equations, Opuscula Mathematica, **32**(2012), 67–74.

[6] L. Cadariu, V. Radu, Fixed points and the stability of Jensen's functional equation, J. Ineq. Pure Appl. Math., 4- ID 4 (2003).

[7] P. Gavruta, S. M. Jung, Y. Li, Hyers-Ulam stability for second order linear differential equations with boundary conditions, Elec. J. Diff. Equ., **80**(2011), 1–5.

[8] P. Gavruta, A generalization of the Hyers-Ulam-Rassias stability of approximately additive mappings, J. Math. Anal. Appl., **184**(1994), 431–436.

[9] M. B. Ghaemi, M. E. Gordji, B. Alizadeh, C. Park, Hyers-Ulam stability of exact second order linear differential equations, Adv. Diff. Equ., **2012**(2012), 1–7.

[10] D. H. Hyers, On the stability of the linear functional equation, Proc. Natl. Acad. Sci., **27**(1941), 222–224.

[11] D. H. Hyers, G. Isac, T. M. Rassias, On the asymptoticity aspect of Hyers -Ulam stability of mappings, Proc. Amer. Math. Soc., **126**(1998), 420–425.

[12] K. W. Jun, D. S. Shin, B. D. Kim, On Hyers-Ulam-Rassias stability of the Pexider equation, J. Math. Anal. Appl., **239**(1999), 20–29.

[13] K. W. Jun, Y. H. Lee, A generalization of the Hyers-Ulam-Rassias stability of the Jenson's equation, J. Math. Anal. Appl., **238**(1999), 305–315.

[14] A. Javedian, E. Sorouri, G. H. Kim, M. E. Gordji, Generalized Hyers-Ulam stability of the second order linear differential equations, J. Appl. Math., **10**(2011), 1–10.

[15] K. W. Jun. S. M. Jung, Y. H. Lee, A generalization of the Hyers-Ulam-Rassias stability of a functional equation of davison, J. Korean Math. Soc., **41**(2004), 501–511.

[16] S. M. Jung, Hyers-Ulam-Rassias stability of Jensen's equation and its application, Proc. Amer. Math. Soc., **126**(1998), 3137–3143.

[17] S. M. Jung, Hyers-Ulam stability of a linear differential equations of first order, Appl. Math. Lett., **17**(2004), 1135–1140.

[18] S. M. Jung, Hyers-Ulam stability of linear differential equations of first order, III, J. Math. Anal. Appl., **311**(2005), 139–146.

[19] S. M. Jung, Hyers-Ulam stability of linear differential equations of first order, II, Appl. Math. Lett., **19**(2006), 854–858.

[20] S. M. Jung, K. S. Lee, Hyers-Ulam-Rassias stability of a linear differential equation, Int. J. Pure Appl. Math., **29**(2006), 107–117.

[21] S. M. Jung, Th. M. Rassias, Ulam's problem for approximate homomorphisms in connection with Bernoulli's differential equation, Appl. Math. Comp., **187**(2007), 223–227.

[22] S. M. Jung, Th. M. Rassias, Generalized Hyers-Ulam stability of Riccati differential equation, Math. Inequ. Appl., **11**(2008), 777–782.

[23] S. M. Jung, J. Brzdek, Hyers-Ulam stability of the delay differential equation $y^{'}(t) = \lambda y(t - \tau)$, Abs. Appl. Anal., **2010**(2010), 1–10.

[24] S. M. Jung, A fixed point approach to the stability of differential equations $y' = F(x, y)$, Bull. Malays. Math. Sci. Soc. **33**(1):(2010), 47–56.

[25] S. M. Jung, *Hyers-Ulam-Rassias Stability of Functional Equations in Nonlinear Analysis*, Springer, 2010.

[26] X. Jiaming, Hyers-Ulam stability of linear differential equations of second order with constant coefficient, Italian J. Pure Appl. Math., **32**(2014), 419–424.

[27] Z. Kominek, On the Hyers-Ulam stability of Pexider-type extension of the Jensen-Hosszu equation, Bull. Int. Math. Virtual Institute, **1**(2011), 53–57.

[28] E. Kreyszig, *Introductory Functional Analysis with Applications*, Wiley, New York, 1989.

[29] Y. Li, Y. Shen, Hyers-Ulam stability of nonhomogeneous linear differential equations of second order, Int. J. Math. Math. Sci.,(2009), 1–7.

[30] Y. Li, Y. Shen, Hyers-Ulam stability of linear differential equations of second order, Appl. Math. Lett., **23**(2010), 306–309.

[31] Y. Li, Hyers-Ulam stability of linear differential equations $y'' = \lambda^2 y$, Thai J. Math., **8**(2010), 215–219.

[32] Y. Li, J. Huang, Hyers-Ulam stability of linear second order differential equations in complex Banach space, Elec. J. Diff. Equ., **184**(2013), 1–7.

[33] Y. H. Lee, K. W. Jun, A generalization of the Hyers-Ulam-Rassias stability of Jensen's equation, J. Math. Anal. Appl., **238**(1999), 305–315.

[34] N. Q. Maher, Hyers-Ulam stability of linear and nonlinear differential equations of second order, Int.J. Appl. Math. Res., **1**(4): (2012), 422–432.

[35] T. Miura, S. E. Takahasi, H. Choda, On the Hyers-Ulam stability of real continuous function valued differentiable map, Tokyo J. Math. **24**(2001), 467–476.

[36] T. Miura, On the Hyers-Ulam stability of a differentiable map, Sci. Math. Japan, **55**(2002), 17–24.

[37] T. Miura, S. Miyajima, S. E. Takahasi, A characterization of Hyers-Ulam stability of first order linear differential operators, J. Math. Anal. Appl., **286**(2003), 136–146.

[38] T. Miura, S. M. Jung, S. E. Takahasi, Hyers-Ulam-Rassias stability of the Banach Space valued linear differential equation $y' = \lambda y$, J. Kor. Math. Soc., **41**(2004), 995–1005.

[39] T. Miura, H. Oka, S. E. Takahasi, N. Niwa, Hyers-Ulam stability of the first order linear differential equation for Banach space–valued holomorphic mappings, J. Math. Ineq., **1**(3):(2007), 377–385.

[40] M. I. Modebei, O. O. Olaiya, I. Otaide, Generalized Hyers-Ulam stability of second order linear ordinary differential equation with initial condition, Adv. Ineq. Appl., (2014), 1–7.

[41] S. M. Moslehian, On the stability of the orthogonal Pexiderized Cauchy equation, J. Math. Anal. Appl., **318**(2006), 211–223.

[42] M. Obloza, Hyers stability of the linear differential equation, Rocznik Naukowo-Dydaktyczny. Prace Matematyczne, **13**(1993), 259–270.

[43] M. Obloza, Connections between Hyers and Lyapunov stability of the ordinary differential equations, Rocznik Naukowo-Dydaktyczny. Prace Matematyczne, **14**(1997), 141–146.

[44] J. A. Oguntuase, On an inequality of Gronwall, J. Ineq. Pure Appl. Math., **2**(2001), 1–6.

[45] C. G. Park, On the stability of the linear mapping in Banach modules, J. Math. Anal. Appl., **275** (2002), 711–720.

[46] C. Park, T. M. Rassias, Fixed points and generalized Hyers-Ulam stability of quadratic functional equation, J. Math. Ineq., **1**(2007), 515–528.

[47] D. Popa, G. Pugna, Hyers-Ulam stability of Euler's differential equation, Results Math., (2015), 1–9.

[48] M. N. Qarawani, Hyers-Ulam stability of linear and nonlinear differential equations of second order, Int. J. Appl. Math. Res., **1**(4):(2012), 422–432.

[49] I. A. Rus, Ulam Stability of ordinary differential equations, Studia Univ. Babes-Bolyai, Mathematica, LIV-4(2009), 125–133.

[50] A. Rahimi, S. Najafzadeh, Hyers-Ulam-Rassias stability of additive type functional equation, General Mathematics, **17**(2009), 45–55.

[51] T. M. Rassias, On the stability of the linear mapping in Banach spaces, Proc. Amer. Math. Soc., **72**(1978), 297–300.

[52] T. M. Rassias, On the stability of functional equations and a problem of Ulam, Acta Appl. Math., **62**(2000), 23–30.

[53] H. Rezaei, S. M. Jung, T. M. Rassias, Laplace transformation and Hyers-Ulam stability of linear differential equations, J. Math. Anal. Appl., **403**(2013), 244–251.

[54] P. K. Sahoo, S. M. Jung, M. S. Moslehian, Stability of a generalized Jensen equation on restricted domains, J. Math. Ineq., **4**(2010), 191–206.

[55] P. K. Sahoo, K. Palaniappan, *Introduction to Functional Equations*, CRC press, New York, 2011.

[56] S. E. Takahasi, T. Miura, S. Miyajima, On the Hyers-Ulam stability of the Banach space valued differential equation $y' = \lambda y$, Bull. Korean Math. Soc., **39**(2002), 309–315.

[57] S. E. Takahasi, H. Takagi, T. Miura, S.Miyajima, The Hyers-Ulam stability constants of first order linear differential operators, J. Math. Anal. Appl., 296(2004), 403–409.

[58] A. K. Tripathy, A. Satapathy, Hyers-Ulam stability of third order Euler's differential equations, J. Nonlinear Dyn., **2014**, ID-487257, 1–6.

[59] A. K. Tripathy, A. Satapathy, Hyers-Ulam stability of fourth order Euler's differential equations, J. Compu. Sci. Appl. Math. 1(2015), 49–58.

[60] A. K. Tripathy, Hyers-Ulam stability of second order linear difference equations, Int. J. Diff. Equs. Appl., 1(2017), 53-65.

[61] A. K. Tripathy, P. Senapati, Hyers-Ulam stability of first order linear difference operators on Banach space, J. Adv. Math., 14(2018), 1–11.

[62] S. M. Ulam, *Problems in Modern Mathematics,* Chapter VI, Science ed. Wiley, New York, 1940.

[63] S. M. Ulam, *A collection of the Mathematical Problems*, Inter. Science Publ., New York, 1960.

[64] S. M. Ulam, *Problems in Modern Mathematics,* Chapter VI, Wiley, New York, 1964.

[65] G. Wang, M. Zhou, L. Sun, Hyers-Ulam stability of linear differential equations of first order, Appl. Math. Lett., **21**(2008), 1024–1028.

[66] Y. Li, Y. Shen, Hyers-Ulam stability of linear differential equations of second order, Appl. Math. Lett., **23**(2010), 306–309.

[67] T. W. Zheng, Z. X. Gao, H. X. Cao, L. Xu, Generalized Hyers-Ulam-Rassias stability of functional inequalities and functional equations, J. Math. Ine., **3**(2009), 63–77.

Index